REA's Test Prep Books Are The Best!

(a sample of the <u>hundreds of letters</u> REA receives each year)

" I did well because of your wonderful prep books... I just wanted to thank you for helping me prepare for these tests. "

Student, San Diego, CA

" My students report your chapters of review as the most valuable single resource they used for review and preparation. "

Teacher, American Fork, UT

" Your book was such a better value and was so much more complete than anything your competition has produced — and I have them all! "

Teacher, Virginia Beach, VA

" Compared to the other books that my fellow students had, your book was the most useful in helping me get a great score. "

Student, North Hollywood, CA

" Your book was responsible for my success on the exam, which helped me get into the college of my choice... I will look for REA the next time I need help. "

Student, Chesterfield, MO

" Just a short note to say thanks for the great support your book gave me in helping me pass the test... I'm on my way to a B.S. degree because of you! "

Student, Orlando, FL

(more on next page)

(continued from front page)

" I just wanted to thank you for helping me get a great score on the AP U.S. History exam... Thank you for making great test preps! "
Student, Los Angeles, CA

" Your *Fundamentals of Engineering Exam* book was the absolute best preparation I could have had for the exam, and it is one of the major reasons I did so well and passed the FE on my first try. "
Student, Sweetwater, TN

" I used your book to prepare for the test and found that the advice and the sample tests were highly relevant... Without using any other material, I earned very high scores and will be going to the graduate school of my choice. "
Student, New Orleans, LA

" What I found in your book was a wealth of information sufficient to shore up my basic skills in math and verbal... The practice tests were challenging and the answer explanations most helpful. It certainly is the *Best Test Prep for the GRE!* "

Student, Pullman, WA

" I really appreciate the help from your excellent book. Please keep up the great work. "
Student, Albuquerque, NM

" I am writing to thank you for your test preparation... your book helped me immeasurably and I have nothing but praise for your *GRE* preparation."
Student, Benton Harbor, MI

(more on back page)

The Best Study Series For
GED
Mathematics
3rd Edition

Michael W. Lanstrum, B.S.
Department of Applied Mathematics
Kent State University, Warren, OH

and

Mel H. Friedman, M.S.
Mathematics Consultant
Plainsboro, NJ

Research & Education Association
Visit our website at
www.rea.com

Research & Education Association

61 Ethel Road West
Piscataway, New Jersey 08854
E-mail: info@rea.com

The Best Study Series for
GED MATHEMATICS

Year 2006 Printing

Printed in the United States of America

Library of Congress Control Number 2002111912

International Standard Book Number 0-87891-433-1

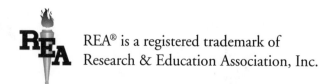

TABLE OF CONTENTS

Introduction

ABOUT THIS BOOK

This book is designed to help you strengthen the skills you will need to take the Mathematics test of the **General Educational Development (GED) Examination**.

A **Pre-Test** section in the beginning of this book will help you assess the areas where you need to work the hardest. After you have completed the review areas and answered all of the drill questions, you will be given a **Post-Test** that will show your improvement in certain areas and show you which areas you still need to study. In the Post-Test section you will answer questions very similar to those you will face on the actual GED.

The reviews cover all areas tested on the Mathematics test of the GED examination. Sections on Whole Numbers, Decimals, Fractions, Percents, Algebra, and other areas of mathematics are covered extensively. Each section contains a drill so you can monitor your progress as you use this book. By mastering the skills presented in this book, you will be able to approach mathematics with confidence.

ABOUT THE GED

The GED is an examination for adults who did not complete high school and would like to earn a high school equivalency diploma. The exam is given by each state, which then issues a GED diploma. The GED is taken by adults who want or need a diploma for work, college, or just personal satisfaction. Nearly 800,000 people take the GED each year.

The GED is broken into five tests: **Language Arts, Writing; Social Studies; Science; Language Arts, Reading; and Mathematics**.

You are given about seven and a half hours to complete the examination. There are a total of 240 multiple-choice questions and an essay question on the GED examination.

The Mathematics test on the GED contains 40 multiple-choice and 10 alternate format questions to be answered in 90 minutes. This test consists of 50 percent arithmetic questions, 30 percent algebra questions, and 20 percent geometry questions. A calculator will be provided for the first half of the test only.

The GED examination is administered by the GED Testing Service of the American Council on Education (ACE) and is developed by writers who have secondary and adult education experience. Because the GED test-takers come from such diverse backgrounds, the ACE makes sure the test writers are also diverse. Once the questions have been written, they are standardized according to a certain level of difficulty and content.

The GED comes in several versions to fit the special needs of its examinees. For example, there are Spanish, French, and Braille versions of the exam, as well as large-print and audio versions.

If you would like to obtain more information about the GED, such as when and where it is administered, contact your local high school or adult education center. You can also get in touch with the GED Testing Service at:

General Educational Development
GED Testing Service
American Council on Education
One Dupont Circle, NW
Washington, D.C. 20036
1-800-62MYGED
(1-800-626-9433)
Website: www.gedtest.org

HOW TO USE THIS BOOK

Before you begin reading the chapters on mathematics, take the Pre-Test at the front of this book. This Pre-Test will assess your current skills and indicate both your strengths and weaknesses.

This book is broken down into seven sections, each dealing with a specific area of mathematics. Each section includes exercises to help you develop your skills as well as skill-building practice exercises at the end of each section to reinforce what you have learned.

It is best to move through this book from front to back because the Introduction to Whole Numbers section lays the foundation for the rest of the book.

Because understanding mathematics is a skill that is developed and not simply a list of formulas to learn, it is important to begin developing this skill as soon as possible. Cramming for this test on the GED simply will not help. It is also wise not to try to do *too* much in one study session. There is a large amount of information presented in this book, and it will take time to digest.

When you are finished with the review sections, take the Post-Test. Compare your Post-Test score with your Pre-Test score and see how much you have improved. You may even want to take the Pre-Test again to re-evaluate your skills.

Before you begin the Pre-Test, you should take inventory of your study skills. Under what circumstances do you study best? Are you most awake in the morning, late afternoon, or evening? Do you study best under bright light or soft? With music, or in total silence? Answer these questions and then try to optimize your study time by creating the ideal learning conditions (the atmosphere in which you learn most efficiently). Also, you should set a spe-cific schedule for yourself that takes into consideration your other commitments. Can you study one hour each morning? A half hour every night? Two or three hours on weekends? Pick a routine and stick to it so you will be confident when it's time to take the exam. Although you are encouraged to write in the margins of this book, you should know that you will not be allowed to mark up the texts on the actual GED examination. When you take the Pre-Test and Post-Test, use a piece of scrap paper instead of writing in the margins of the book. This will help you be more comfortable with the actual test format.

Please note that the GED won't penalize you for guessing wrong, so if you are stuck on a question, don't leave it blank. Instead, eliminate any answers you know are not correct and choose one of the remaining options. You have a better chance of getting a question right by making an educated guess, and then you can move on to the next question. Remember, the exam is timed, so you do not want to spend too much time on any one question.

ABOUT RESEARCH & EDUCATION ASSOCIATION

Founded in 1959, Research & Education Association is dedicated to publishing the finest and most effective educational materials—including software, study guides, and test preps—for students in middle school, high school, college, graduate school, and beyond.

REA's Test Preparation series includes books and software for all academic levels in almost all disciplines. Research & Education Association publishes test preps for students who have not yet completed high school, as well as high school students preparing to enter college. Students from countries around the world seeking to attend college in the United States will find the assistance they need in REA's publications. For college students seeking advanced degrees, REA publishes test preps for many major graduate school admission examinations in a wide variety of disciplines, including engineering, law, and medicine. Students at every level, in every field, with every ambition can find what they are looking for among REA's publications.

REA's practice tests are always based upon the most recently administered exams, and include every type of question that you can expect on the actual exams.

REA's publications and educational materials are highly regarded for their significant contribution to the quest for excellence that characterizes today's educational goals. We continually receive an unprecedented amount of praise from professionals, instructors, librarians, parents, and students for our books. Our authors are as diverse as the subjects represented in the books we publish. They are well-known in their respective fields and serve on the faculties of prestigious high schools, colleges, and universities throughout the United States and Canada.

Today, REA's wide-ranging catalog is a leading resource for teachers, students, and professionals.

We invite you to visit us at *www.rea.com* to find out how "REA is making the world smarter."

ACKNOWLEDGMENTS

Special recognition is extended to the following persons:

Pam Weston, VP, Publishing, for setting the quality standards for production integrity and managing the publication to completion.

Larry B. Kling, VP, Editorial, for supervising revisions.

Robert Herrmann, Anne Reifsnyder, and Ellen Gong for their editorial contributions.

Jeanne Audino, Senior Editor, for preflight editorial review.

Diane Goldschmidt, Associate Editor, for post-production quality assurance.

Network Typesetting for typesetting the manuscripts.

Mathematics

Pre-Test

MATHEMATICS

PRE-TEST

DIRECTIONS: Choose the best answer choice for each question below.

1. What is the value of the digit 3 in the number 735,246?

   (1) Hundreds

   (2) Thousands

   (3) Ten thousands

2. 718 _____ 781. Which of the following is the correct symbol between the numbers?

   (1) <

   (2) >

   (3) =

3. What is the number 2,536 rounded off to the underlined digit?

   (1) 2,500

   (2) 2,540

   (3) 2,600

4. What is the sum of y^2 and $5y^2$?

   (1) $5y^4$

   (2) $6y^4$

   (3) $6y^2$

5. What is the sum of 3.156, 5.6, and 2.27?

   (1) 11.026

   (2) 34.39

   (3) 343.9

6. What is the sum of $\frac{5}{6}$ and $\frac{1}{3}$?

   (1) $\frac{2}{3}$

   (2) $\frac{7}{9}$

   (3) $1\frac{1}{6}$

7. What is the value of $6 - 3\frac{5}{12}$?

   (1) $3\frac{5}{12}$

   (2) $3\frac{7}{12}$

   (3) $2\frac{7}{12}$

3

8. What is the value of $5\frac{1}{4} \times 1\frac{1}{7}$?

(1) $5\frac{1}{28}$

(2) 6

(3) $6\frac{11}{28}$

9. What is the value of $9 \times 8 - 2 \times 5$?

(1) 62

(2) 270

(3) 350

10. What is 129% written as a decimal?

(1) 129.0

(2) 12.9

(3) 1.29

11. What is $4x \times -3xy$?

(1) $-12x^2y$

(2) $-7x^2y$

(3) $-xy$

12. What is the square root of 64?

(1) 4,096

(2) 32

(3) 8

13. Five percent of what number is 15?

(1) 0.75

(2) 33.33

(3) 300

14. What is the average of 5, 8, 9, and 10?

(1) 8.5

(2) 8.0

(3) 6.4

15. Four pieces of paper numbered 1, 2, 4, and 8 are placed in a can and one is drawn without looking. What is the probability of drawing an even number?

(1) $\frac{3}{4}$

(2) $\frac{1}{2}$

(3) $\frac{1}{3}$

16. A die is rolled on a table. What is the probability of rolling a 4?

(1) $\frac{1}{6}$

(2) $\frac{1}{4}$

(3) $\frac{1}{2}$

17. Simplify the following:

$3ac^2 + 2b^2c + 7ac^2 + 2a^2 + b^2c =$

(1) $12ac^2 + 3b^2c$

(2) $11ac^2 + 4ab^2c$

(3) $15a^2b^4c^4$

18. A $580 TV is going on sale for 30% off. What will be the discounted price?

    (1) $550

    (2) $406

    (3) $174

19. Simplify the following: $6m^6 \div 2m^2$.

    (1) $4m^4$

    (2) $3m^4$

    (3) $3m^3$

20. What is the area of a square with a side of 6 inches?

    (1) 12 inches^2

    (2) 24 inches^2

    (3) 36 inches^2

21. What is the approximate circumference of a circle with a radius of 5 inches?

    (1) 15.7 inches

    (2) 31.4 inches

    (3) 78.5 inches

22. What is the approximate area of a circle with a diameter of 8 inches?

    (1) 50.3 inches^2

    (2) 201.1 inches^2

    (3) 631.7 inches^2

23. What is the number 26,481 rounded off to the nearest hundred?

    (1) 26,000

    (2) 26,400

    (3) 26,500

24. 4,356 _____ 4,536. Which of the following is the correct symbol between these numbers?

    (1) >

    (2) =

    (3) <

25. For the equation $\frac{1}{3}x = 9$, which of the following is true?

    (1) $x = 3$

    (2) $x = 8\frac{2}{3}$

    (3) $x = 27$

26. What is the value of 1.4×0.027?

    (1) 0.00378

    (2) 0.0378

    (3) 0.378

27. What is the value of $7\frac{1}{3} - 2\frac{3}{5}$?

    (1) $5\frac{4}{15}$

    (2) $4\frac{11}{15}$

    (3) $4\frac{4}{15}$

28. Five numbers have a mean of 20. If four of these numbers are 12, 14, 17, and 34, what is the fifth number?

    (1) 21

    (2) 22

    (3) 23

29. What is the value of $(7 + 8 \times 3) \div 4$?

    (1) 13

    (2) $11\frac{1}{4}$

    (3) $7\frac{3}{4}$

30. What is 0.357 written as a percent?

    (1) 0.00357%

    (2) 35.7%

    (3) 357%

31. What is 4.18% written as a decimal?

    (1) 0.418

    (2) 0.0418

    (3) 0.00418

32. What is 16% of 40?

    (1) 0.004

    (2) 6.4

    (3) 250

33. What percent of 20 is 25?

    (1) 20%

    (2) 80%

    (3) 125%

34. Which one of the following could represent the three sides of an equilateral triangle?

    (1) 7, 8, 9

    (2) 7, 8, 7

    (3) 7, 7, 7

35. Five pieces of paper numbered 2, 4, 7, 9, and 12 are placed in a jar and one number is drawn without looking. What is the probability of drawing a number greater than 8?

    (1) $\frac{1}{5}$

    (2) $\frac{2}{5}$

    (3) $\frac{1}{2}$

36. An ordinary die is rolled. What is the probability of getting a 6?

    (1) $\frac{1}{6}$

    (2) $\frac{1}{2}$

    (3) $\frac{5}{6}$

37. Solve for x: $-6x + 9 < -5x$

    (1) $x > 9$

    (2) $x > -9$

    (3) $x < 9$

38. A $140 radio is being sold at a 40% discount. What is the discounted price?

    (1) $56

    (2) $70

    (3) $84

39. A rectangle has a length of $10^1/_3$ and a width of 6. What is the perimeter?

    (1) $26\frac{1}{3}$

    (2) $32\frac{2}{3}$

    (3) 62

40. The perimeter of a square is 25. What is the length of one side?

    (1) 5

    (2) $6\frac{1}{4}$

    (3) $12\frac{1}{2}$

41. The length of a rectangle is 14 and the area is 49. What is the width?

    (1) $10\frac{1}{2}$

    (2) 7

    (3) $3\frac{1}{2}$

42. What is the approximate area of a circle with a radius of 4.5?

    (1) 85.9

    (2) 63.6

    (3) 28.3

43. What is the approximate circumference of a circle with a diameter of 9.6?

    (1) 30.2

    (2) 45.7

    (3) 60.3

44. What is the average of 92, 85, 76, and 47?

    (1) 85

    (2) 75

    (3) 133

45. Which of the following lists of numbers has two modes?

    (1) 1, 1, 1, 2, 2, 3, 3, 3, 4

    (2) 1, 1, 1, 2, 2, 2, 2, 3, 3, 3, 4

    (3) 1, 1, 1, 1, 2, 2, 2, 3, 3, 4

46. Which set of numbers are the length of the sides of an equilateral triangle?

    (1) 4, 4, 8

    (2) 9, 7, 2

    (3) 8, 8, 8

47. What is 92,842 rounded to the tens position?

    (1) 92,843

    (2) 92,852

    (3) 92,840

48. What is the sum of 4.82, 19.57, and 42.030?

    (1) 68.28

    (2) 66.42

    (3) 6.642

49. The pie graph below shows amounts of money spent by government agencies, in millions of dollars, during a 3-year period. Approximately what percent of the total money did NASA spend?

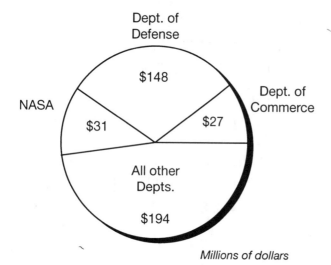

Millions of dollars

(1) 8%

(2) 12%

(3) 16%

50. Solve for x: $5x - 3 > 17$

(1) $x < 15$

(2) $x > \dfrac{14}{5}$

(3) $x > 4$

For problems 51–53, choose the best method for setting up the problem.

51. What is 60% of 50?

(1) $60 \div 50$

(2) 0.60×50

(3) 60×0.50

52. What is the average of 9, 10, 10, 20, and 16?

(1) $9 + 10 + 10 + 20 + 16$

(2) $(9 + 10 + 10 + 20 + 16) \times 5$

(3) $(9 + 10 + 10 + 20 + 16) \div 5$

53. A rectangular garden has a length of $7^{1}/_{2}$ feet and a width of $3^{1}/_{2}$ feet. What is the perimeter?

(1) $(2 \times 7\frac{1}{2}) + (2 \times 3\frac{1}{2})$

(2) $(2 \times 7\frac{1}{2}) + (3\frac{1}{2} \div 2)$

(3) $7\frac{1}{2} \times 3\frac{1}{2}$

54. $3x + 5x - x =$ _____

55. If $2x + 7 = 4x + 2$, then

$x =$ _____

56. What are the coordinates of point p?

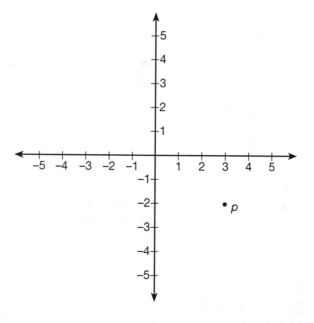

MATHEMATICS

ANSWER KEY

1. (3)	15. (1)	29. (3)	43. (1)
2. (1)	16. (1)	30. (2)	44. (2)
3. (1)	17. (1)	31. (2)	45. (1)
4. (3)	18. (2)	32. (2)	46. (3)
5. (1)	19. (2)	33. (3)	47. (3)
6. (3)	20. (3)	34. (3)	48. (2)
7. (3)	21. (2)	35. (2)	49. (1)
8. (2)	22. (1)	36. (1)	50. (3)
9. (1)	23. (3)	37. (1)	51. (2)
10. (3)	24. (3)	38. (3)	52. (3)
11. (1)	25. (3)	39. (2)	53. (1)
12. (3)	26. (2)	40. (2)	54. $7x$
13. (3)	27. (2)	41. (3)	55. $x = \dfrac{5}{2}$
14. (2)	28. (3)	42. (2)	56. $(3, -2)$

PRE-TEST SELF-EVALUATION

Question Number	Subject Matter Tested	Section to Study (section, heading)
1.	Place Value	Whole Numbers, Introduction to Whole Numbers
2.	Whole Numbers	Whole Numbers, Introduction to Whole Numbers
3.	Rounding Numbers	Whole Numbers, Introduction to Whole Numbers
4.	Algebra	Algebra, Operations with Polynomials—Addition
5.	Decimals	Decimals, Addition of Decimals
6.	Fractions	Fractions, Addition of Fractions
7.	Fractions	Fractions, Subtraction of Fractions
8.	Fractions	Fractions, Multiplication of Fractions
9.	Order of Operations	Topics in Mathematics, Order of Operations
10.	Percents	Percents, Introduction to Percents
11.	Algebra	Algebra, Operations with Polynomials—Multiplication
12.	Square Roots	Topics in Mathematics, Square Roots
13.	Percents	Percents, Percent Problems—Working with Percents
14.	Averages	Topics in Mathematics, Averages
15.	Probability	Advanced Topics, Probability
16.	Probability	Advanced Topics, Probability
17.	Algebra	Algebra, Operations with Polynomials—Addition
18.	Percents	Percents, Percent Problems—Working with Percents
19.	Algebra	Algebra, Operations with Polynomials—Division
20.	Area	Topics in Mathematics, Measurement
21.	Circumference	Advanced Topics, Geometry
22.	Area	Topics in Mathematics, Measurement
23.	Rounding Numbers	Whole Numbers, Introduction to Whole Numbers
24.	Place Value	Whole Numbers, Introduction to Whole Numbers
25.	Algebra	Algebra, Equations
26.	Decimals	Decimals, Multiplication of Decimals
27.	Fractions	Fractions, Subtraction of Fractions
28.	Statistics	Advanced Topics, Statistics
29.	Order of Operations	Topics in Mathematics, Order of Operations

Question Number	Subject Matter Tested	Section to Study (section, heading)
30.	Percents	Percents, Percent Problems—Introduction to Percents
31	Percents	Percents, Introduction to Percents
32.	Percents	Percents, Percent Problems—Working with Percents
33.	Percents	Percents, Percent Problems—Working with Percents
34.	Triangles	Advanced Topics, Geometry
35.	Probability	Advanced Topics, Probability
36.	Probability	Advanced Topics, Probability
37.	Algebra	Algebra, Inequalities
38.	Percents	Percents, Percent Problems—Working with Percents
39.	Perimeter	Topics in Mathematics, Measurement
40.	Perimeter	Topics in Mathematics, Measurement
41.	Area	Topics in Mathematics, Geometry
42.	Area	Topics in Mathematics, Geometry
43.	Circles	Topics in Mathematics, Geometry
44.	Averages	Topics in Mathematics, Averages
45.	Statistics	Advanced Topics, Statistics
46.	Triangles	Advanced Topics, Geometry
47.	Place Value	Whole Numbers, Introduction to Whole Numbers
48.	Decimals	Decimals, Addition of Decimals
49.	Data Analysis	Advanced Topics, Data Analysis
50.	Algebra	Algebra, Inequalities
51.	Percents	Percents, Percent Problems—Working with Percents
52.	Averages	Topics in Mathematics, Averages
53.	Perimeter	Topics in Mathematics, Measurement
54.	Algebra	Algebra, Operations with Polynomials-Addition and Subtraction
55.	Algebra	Algebra, Equations
56.	Graphs	Advanced Topics, Graphs

PRE-TEST
ANSWERS AND EXPLANATIONS

1. **(3)** The fifth place to the left of the right-most digit represents the ten thousands place.

2. **(1)** From left to right for two numbers with the same number of placeholders, the first difference occurs in the tens place, where 1 in 718 is less than 8 in 781.

3. **(1)** Since the tens digit 3 is less than 5, leave the 5 as is and change the 3 and 6 to 0's.

4. **(3)** $y^2 + 5y^2 = (1+5)(y^2) = 6y^2$

5. **(1)** 11.026

$$
\begin{array}{r}
\overset{1}{}\overset{1}{} \\
3.156 \\
5.6 \\
+\ 2.27 \\
\hline
11.026
\end{array}
$$

6. **(3)**

$$\frac{5}{6} + \frac{1}{3} = \frac{5}{6} + \frac{2}{6} = \frac{7}{6} = 1\frac{1}{6}$$

7. **(3)**

$$6 - 3\frac{5}{12} = \frac{6}{1} - \frac{41}{12} = \frac{72}{12} - \frac{41}{12} = \frac{31}{12} = 2\frac{7}{12}$$

8. **(2)**

$$5\frac{1}{4} \times 1\frac{1}{7} = \frac{21}{4} \times \frac{8}{7} = \frac{168}{28} = 6$$

9. **(1)**

$$9 \times 8 - 2 \times 5 = 72 - 10 = 62$$

10. **(3)** To convert a percent to a decimal, move the decimal point two places to the left. 129% = 1.29

11. **(1)** Multiply the coefficients but add the exponents of similar variables.

$$4x - 3 = -12$$

$$x^1 \times x^1 = x^2$$

y remains unchanged so

$$(4x)(-3xy) = -12x^2y$$

12. **(3)**

$$\sqrt{64} = 8, \text{ since } 8 \times 8 = 64$$

13. **(3)**

$$1.5 \div 5\frac{1}{2} = \frac{15}{0.25} = 300$$

14. **(2)**

$$(5 + 8 + 9 + 10) \div 4 = 32 \div 4 = 8.0$$

15. **(1)** There are three even numbers out of a total of four numbers, so the probability is $^3/_4$.

16. **(1)** There is only one 4 out of a total of six numbers, so the probability is $^1/_6$.

17. **(1)** Group like terms and combine to get

$$3ac^2 + 7ac^2 + 2ac^2 = 12ac^2$$

$$2b^2c + b^2c = 3b^2c$$

Add to get $12ac^2 + 3b^2c$

18. **(2)**

$$(\$580) \times (0.30) = \$174$$

Then

$$\$580 - \$174 = \$406$$

19. **(2)** Divide the coefficients but subtract the exponents of similar variables

$$6 \div 2 = 3$$

$$m^6 \div m^2 = m^4$$

The answer is $3m^4$

20. **(3)**

Area $= 6^2 = 36$ inches^2

21. **(2)**

Circumference $= (2\pi) \times (5) = 10\pi$

$= 31.4$ inches

22. **(1)**

Area of a circle $= (\pi) \times (\text{radius})^2$

$= (\pi) \times (4)^2 = \pi \times 16 = 50.3$ inches^2

23. **(3)** Since the tens digit 8 is at least 5, raise the hundreds digit 4 to a 5. All digits to its right become zeros.

24. **(3)** The hundreds place is the first instance where the numbers differ. When reading from left to right, the 3 in 4,356 is less than the 5 in 4,536.

25. **(3)** Divide both sides of the equation by (1/3) so

$$\frac{1}{3}x \div \frac{1}{3} = 9 \div \frac{1}{3}$$

$$x = 9 \times 3 = 27$$

26. **(2)** 0.0378

$$\begin{array}{c} 1.4 \\ \times\, 0.027 \end{array} \longrightarrow \begin{array}{c} 14 \\ \times\, 0027 \\ \hline 98 \end{array} \longrightarrow \begin{array}{c} 14 \\ \times\, 0027 \\ \hline 98 \\ +\, 280 \\ \hline \end{array} \longrightarrow$$

$$\begin{array}{c} 14 \\ \times\, 0027 \\ \hline 98 \\ \times\, 280 \\ \hline 378 \end{array} \longrightarrow$$

1.4 (1 decimal place)
$\times\, 0.027$ (3 decimal places)
0.0378 (4 decimal places)

27. **(2)**

$$7\frac{1}{3} - 2\frac{3}{5} = \frac{22}{3} - \frac{13}{5} = \frac{110}{15} - \frac{39}{15}$$

$$= \frac{71}{15} = 4\frac{11}{15}$$

28. **(3)** The sum of the five numbers must be $5 \times 20 = 100$. Then the missing fifth number is $100 - (12 + 14 + 17 + 34) = 23$.

29. **(3)**

$$(7 + 8 \times 3) \div 4 = (7 + 24) \div 4$$

$$= 31 \div 4 = 7\frac{3}{4}$$

30. **(2)** To change a decimal to a percent, move the decimal point two places to the right and add a percent sign.

31. **(2)** To change a percent to a decimal, move the decimal point two places to the left and drop the percent sign.

32. **(2)**

$$16\% = 0.16$$

Then

$$0.16 \times 40 = 6.4$$

33. **(3)**

$$\frac{25}{20} = 1.25 = 125\%$$

34. **(3)**   In an equilateral triangle, all sides must be the same length.

35. **(2)**   Since there are two numbers greater than 8, the probability is $^2/_5$.

36. **(1)**   There is only one 6 out of six numbers, so the probability is $^1/_6$.

37. **(1)**   Add $6x$ to both sides of the inequality to get $9 < x$, which is equivalent to $x > 9$.

38. **(3)**

$$\$140 \times 0.40 = \$56.$$

Then

$$\$140 - \$56 = \$84$$

39. **(2)**

$$\text{Perimeter} = 2 \times 10\frac{1}{3} + 2 \times 6 = 32\frac{2}{3}$$

40. **(2)**   Each side

$$= \frac{25}{4} = 6\frac{1}{4}$$

41. **(3)**   The width

$$= \frac{49}{14} = 3\frac{1}{2}$$

42. **(2)**   The area $= \pi \times 4.5^2 = 20.25\pi$, which is approximately 63.6.

43. **(1)**   The circumference $= \pi \times 9.6$, which is approximately 30.2.

44. **(2)**   To determine an average, add all the numbers and divide by the number of numbers. In this case, 92, 85, 76, and 47 equal 300. 300 divided by 4 is 75.

45. **(1)**   Both numbers 1 and 3 occur three times, which is a higher frequency than the other two numbers.

46. **(3)**   In an equilateral triangle, all sides must be of equal length.

47. **(3)**   Since the ones digit is less than 5, leave the 4 and change the 2 to a 0.

48. **(2)**   When adding decimals, line up the decimal points and add as you would add whole numbers.

49. **(1)**   To solve the problem, first add to find the total amount of government spending.

$$31 + 148 + 27 + 194 = 400$$

Then, to find the percent of NASA money spent, divide the total government spending by NASA's spending to get

$$31 \div 400 = 0.0775 = 8\%$$

50. **(3)**

$$5x - 3 + 3 \;\; > 17 + 3$$

$$\frac{5x}{5} \;\; > \;\; \frac{20}{5}$$

$$x > 4$$

51. **(2)** In word problems, "of" indicates multiplication, and percents should be converted into decimals. Therefore, the problem should be set up as

$$0.60 \times 50$$

52. **(3)** To find the average of 5 numbers, add them and divide the total by 5.

$$(9 + 10 + 10 + 20 + 16) \div 5$$

53. **(1)** To determine the perimeter of a rectangle, use the formula

Perimeter = 2 (length) + 2 (width)

$$2 \left(7\frac{1}{2}\right) + 2 \left(3\frac{1}{2}\right)$$

54. Combine like terms by summing the coefficients separate from the variable.

$$3x + 5x - x = (3 + 5 - 1)(x) = 7x$$

55.

$$
\begin{aligned}
2x + 7 - 4x &= 4x + 2 - 4x \\
-2x + 7 - 7 &= 2 - 7 \\
-2x \div -2 &= -5 \div -2 \\
x &= 5/2
\end{aligned}
$$

56. From $(0, 0)$, point p is located 3 units to the right on the x-axis and 2 units down on the y-axis. Thus its coordinates are $(3, -2)$.

Mathematics

Whole Numbers

MATHEMATICS

WHOLE NUMBERS

INTRODUCTION TO WHOLE NUMBERS

Numbers have been around since the beginning of time. Their appearances may have changed throughout the years, but the ideas have not. Cave dwellers used numbers in a manner that was very similar to the way we use them today. Whether they were counting rocks, trees, animals, or whatever was around, numbers played a very important role in their lives.

Numbers can be classified into a variety of types. The most important type is called **whole numbers**. This group (or set) of numbers begins with zero (0) and increases to infinity. Below is a partial list of whole numbers:

0, 1, 2, 3, 4, 5, 6, 7, 8, 9, 10, 11, 12, …

The three dots at the end of this list indicate that this pattern continues on and on (and on). These numbers are used in our everyday life. Some examples would be measuring distances and times, working with money, and mailing a letter (to name a few).

As whole numbers increase, the number of digits also increases. We already know that a three-digit number (such as 321) is larger than a one-digit number (such as 5). Which one is "better" depends on what we are considering. If we use these numbers to compare salaries, we would all prefer to get paid $321. On the other hand, if these numbers were prices of two identical shoes, we would rather pay $5.

Another way of considering smaller and larger numbers is by discussing their **place value**. Consider the following number:

152,738,946

The value of a digit depends on where it appears in the number. As we move from the right to the left, the values of the digits get larger and larger. We have listed the value of each digit in the above number:

1 is in the hundred millions place

5 is in the ten millions place

2 is in the millions place

7 is in the hundred thousands place

3 is in the ten thousands place

8 is in the thousands place

9 is in the hundreds place

4 is in the tens place

6 is in the ones place

This number would be read as "one hundred fifty-two million, seven hundred thirty-eight thousand, nine hundred forty-six."

At first glance, writing out large numbers in words may appear unnecessary, but we do it on a regular basis. Every time we write a check, we write the amount in words. In the following examples, we will consider place values, writing numbers in words, and writing numbers from words.

Example 1

Write the place value of "5" in the following numbers.

(a) 1<u>5</u>2

This "5" is in the <u>tens</u> place.

(b) 6<u>5</u>,207

This "5" is in the <u>thousands</u> place.

(c) 7,892,34<u>5</u>

This "5" is in the <u>ones</u> place.

(d) 3,<u>5</u>00

This "5" is in the <u>hundreds</u> place.

Example 2

Write the following numbers in words.

(a) 975

<u>Nine hundred seventy-five</u>

(b) 24,803

<u>Twenty-four thousand, eight hundred three</u>

(c) 534,600

<u>Five hundred thirty-four thousand, six hundred</u>

(d) 3,002,187

<u>Three million, two thousand, one hundred eighty-seven</u>

Example 3

Write the following numbers using digits.

(a) Twenty-one thousand, four hundred sixty-seven

<u>21,467</u>

(b) Three hundred thousand, five hundred one

<u>300,501</u>

(c) One million, thirty-six thousand, seven hundred eighty

<u>1,036,780</u>

(d) Four hundred fifty-nine

<u>459</u>

There is still another way in which we can compare two numbers. If given two numbers, they must be related in one of the following three ways:

= means *equals*

> means *is greater than*

< means *is less than*

A simple way to remember this is that for the symbols < and >, the arrow points at the smaller number.

Example 4

Compare the following pairs of numbers. Write the appropriate symbol (<, >, =) between the two numbers.

(a) 534 _____ 6,980

We know that 6,980 is a larger number because it has more digits than 534.

(b) 3,158 _____ 3,128

We know that 3,158 is a larger number because, moving from left to right, the first different digit is in the tens position, and 5 is larger than 2.

(c) 62,947 _____ 62,947

We know that these numbers are equal because their corresponding digits are the same.

(d) 7,267 _____ 9,000

We know that 9,000 is a larger number because, moving from left to right, the first different digit is in the thousands position, and 9 is larger than 7.

In some cases, the exact numerical value is not needed. In these situations, an approximation is close enough. An example of this is when counting items—if our cave dweller is out counting trees, do we need to know that there are exactly 32 trees around his "home" or is 30 close enough? Another example is income tax—instead of writing $35,928.33 as our income, we are allowed to "estimate" this value to be $36,000. In these examples, we are actually **rounding** the exact numbers. We must first pick the place we want to round the number. Mark this digit (for example, underline it). Examine the digit to the right of the marked digit. If this digit is a 5 or more, add 1 to the marked digit. If this digit is less than 5, do not change the marked digit. Change all digits to the right of the marked digit to zeros.

Example 5

Round the following numbers to the appropriate position.

(a) Round 4,365 to the tens position.

4,3<u>6</u>5—Since the digit to the right is 5, we would add 1 to the marked digit and change the digits to the right to zeros = 4,3<u>7</u>0.

(b) Round 85,725 to the thousands position.

8<u>5</u>,725—Since the digit to the right is 7, we would add 1 to the marked digit and change the digits to the right to zeros = 8<u>6</u>,000.

(c) Round 631,654 to the hundred thousands position.

<u>6</u>31,654—Since the digit to the right is 3, we would leave the marked digit alone and change the digits to the right to zeros = <u>6</u>00,000.

(d) Round 5,486 to the hundreds position.

5,<u>4</u>86—Since the digit to the right is 8, we would add 1 to the marked digit and change the digits to the right to zeros = 5,<u>5</u>00.

Questions

Write the place value of the underlined digit in the following numbers.

1. 23,8<u>5</u>6   _____

2. 34<u>7</u>,581   _____

3. <u>7</u>,843,165   _____

4. 436,<u>8</u>95   _____

5. <u>5</u>4,165   _____

Write the following numbers in words.

6. 564   _____

7. 347,583   _____

8. 9,340   _____

9. 1,654,777   _____

10. 501,600   _____

Write the following numbers using digits.

11. Seven thousand, eight hundred ninety-seven

_____

12. Fifty-seven thousand, six hundred twenty-one

_____

13. Two hundred seventeen thousand, eight hundred thirty

_____

14. Six hundred twenty-four thousand, eight hundred sixty-six

_____

15. One million, two hundred thousand, three hundred

_____

Compare the following pairs of numbers. Write <, =, or > between the two numbers.

16. 3,499 _____ 3,500

17. 95,456 _____ 4,579

18. 76,854 _____ 76,854

19. 1,287,663 _____ 1,287,563

20. 4,857 _____ 37,875

Round the following numbers to the underlined position.

21. 438,<u>6</u>53 _____

22. <u>5</u>25,376 _____

23. 84,6<u>5</u>1 _____

24. <u>8</u>,456,854 _____

25. 9<u>7</u>,966 _____

Answers

1. Tens. The second digit to the left of the decimal point is the tens place value.

2. Thousands. The fourth digit to the left of the decimal point is the thousands place value.

3. Millions. The seventh digit to the left of the decimal point is the millions place value.

4. Hundreds. The third digit to the left of the decimal point is the hundreds place value.

5. Ten thousands. The fifth digit to the left of the decimal point is the ten thousands place value.

6. Five hundred sixty-four

7. Three hundred forty-seven thousand, five hundred eighty-three

8. Nine thousand, three hundred forty

9. One million, six hundred fifty-four thousand, seven hundred seventy-seven

10. Five hundred one thousand, six hundred

11. 7,897

12. 57,621

13. 217,830

14. 624,866

15. 1,200,300

16. < 3,499 is less than 3,500.

17. > 95,456 is greater than 4,579.

18. = The two numbers are of equal value.

19. > 1,287,663 is greater than 1,287,563.

20. < 4,857 is less than 37,875.

21. 438,<u>7</u>00 When rounding, if the number to the right of the number you are rounding to is 5 or greater, round up.

22. 5̲00,000 If the number next to the number you are rounding to is less than 5, do not round up.

23. 84,65̲0 When rounding, if the number to the right of the number you are rounding to is 5 or greater, round up.

24. 8̲,000,000 If the number next to the number you are rounding to is less than 5, do not round up.

25. 98̲,000 When rounding, if the number to the right of the number you are rounding to is 5 or greater, round up.

ADDITION OF WHOLE NUMBERS

Now that we have a better understanding of what a whole number is, the question becomes:

What do we do with them?

This question has a variety of answers. In this section, we will discuss how we can add these numbers. In following sections, we will address operations such as subtraction, multiplication, and division. As we continue our "career" in mathematics, we will see more and more things that whole numbers can do.

The easiest way to add whole numbers is to write the numbers in a column with the place values lined up. That is, the ones digits are lined up in one column, the tens digits are lined up in the next column, etc. Then we add the digits down each column, moving from the right to the left to calculate our overall answer. If the sum of a column is ten or more, we place the right-hand digit of the column answer in our overall answer and "carry" the tens (or possibly hundreds) position to the next column. In the following examples, we will consider some problems where carrying is necessary.

Example 1

Add the following pairs of numbers.

(a) Add 341 and 657.

$$
\begin{array}{r} 341 \\ + \; 657 \\ \hline \end{array}
\longrightarrow
\begin{array}{r} 341 \\ + \; 657 \\ \hline 998 \end{array}
$$

By adding the ones column, we get $1 + 7 = 8$. The tens column gives $4 + 5 = 9$, and the hundreds column gives $3 + 6 = 9$. Notice that, in this problem, carrying is not needed. The final answer is 998.

(b) Add 8,217 and 340.

$$
\begin{array}{r} 8217 \\ + \; 340 \\ \hline \end{array}
\longrightarrow
\begin{array}{r} 8217 \\ + \; 340 \\ \hline 8557 \end{array}
$$

By adding the ones column, we get $7 + 0 = 7$. The tens column gives $1 + 4 = 5$. The hundreds column gives $2 + 3 = 5$, and the thousands column gives $8 +$ "nothing" (or $8 + 0$), resulting in 8. Again, notice that carrying is not necessary. The final answer is 8,557.

(c) Add $734 + 248$.

$$
\begin{array}{r} 734 \\ + \; 248 \\ \hline \end{array}
\longrightarrow
\begin{array}{r} {}^{1} \\ 734 \\ + \; 248 \\ \hline 982 \end{array}
$$

In this example, carrying is a must. When we add the ones column we get $4 + 8 = 12$. We place the 2 in our answer line and carry the 1 to the tens column. (This is the 1 appearing above the 3 in the rewrite of the problem.) Then, when we add the tens column, we have $1 + 3 + 4 = 8$. We continue to add down the columns from right to left and get 7 + 2 = 9 in the hundreds position in our answer. The final answer is 982.

(d) Add 8,765 and 1,073.

$$
\begin{array}{r}
8765 \\
+\ 1073 \\
\end{array}
\longrightarrow
\begin{array}{r}
1 \\
8765 \\
+\ 1073 \\
\hline
9838 \\
\end{array}
$$

Carrying is also necessary in this example. Adding the ones column gives 5 + 3 = 8. The tens column gives 6 + 7 = 13. Here we place the 3 in the answer line and carry the 1 to the hundreds column. (This is the 1 that appears above the 7 in the rewrite of this problem.) The hundreds column gives 1 + 7 + 0 = 8, and finally, the thousands column gives 8 + 1 = 9. The final answer is 9,838.

Example 2

Add the following numbers.

(a) Add 345, 401, and 232.

$$
\begin{array}{r}
345 \\
401 \\
+\ 232 \\
\end{array}
\longrightarrow
\begin{array}{r}
345 \\
401 \\
+\ 232 \\
\hline
978 \\
\end{array}
$$

By adding the ones column, we get 5 + 1 + 2 = 8. The tens column gives 4 + 0 + 3 = 7, and the hundreds column gives 3 + 4 + 2 = 9. Carrying is not used in this problem. The final answer is 978.

(b) Add 756, 893, and 126.

$$
\begin{array}{r}
756 \\
893 \\
+\ 126 \\
\end{array}
\longrightarrow
\begin{array}{r}
1\ 1 \\
756 \\
893 \\
+\ 126 \\
\hline
1775 \\
\end{array}
$$

By adding the ones column, we get 6 + 3 + 6 = 15. We place the 5 in our answer line and carry the 1 to the tens column. (This is why the 1 appears over the 5 in the rewrite of this problem.) When we add the tens column, we get 1 + 5 + 9 + 2 = 17. The 7 is placed in our answer line and the 1 is carried to the hundreds column. When the hundreds column is added, we get 1 + 7 + 8 + 1 = 17. Since we have no more columns to add, we can write 17 in our answer line. The final answer is 1,775.

(c) Add 6,458, 3,567, and 3,466.

$$
\begin{array}{r}
6458 \\
3567 \\
+\ 3466 \\
\end{array}
\longrightarrow
\begin{array}{r}
1\,1\,2 \\
6458 \\
3567 \\
+\ 3466 \\
\hline
13491 \\
\end{array}
$$

By adding the ones column, we get 8 + 7 + 6 = 21. We can place the 1 in our answer line and carry the 2 to the tens column. (This accounts for the 2 above the 5 in the rewrite of the problem.) In the tens column, we get 2 + 5 + 6 + 6 = 19. We can place the 9 in our answer line and carry the 1 to the hundreds column. (This accounts for the 1 over the 4 in the above example.) The hundreds column gives 1 + 4 + 5 + 4 = 14. The 4 is placed in the answer line and the 1 is carried to the thousands column. (This accounts for the 1 that appears above the 6 in the above problem.) By adding the thousands column, we get 1 + 6 + 3 + 3 = 13. Since we have no more columns, we can write the 13 in our answer line. The final answer is 13,491.

(d)   Add 3,456, 2,589, 3,764, and 4,607.

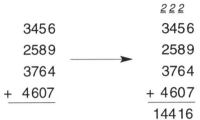

```
          2 2 2
  3456    3456
  2589    2589
  3764    3764
+ 4607  + 4607
         14416
```

By adding the ones column, we get 6 + 9 + 4 + 7 = 26. We can place the 6 in our answer line and carry the 2 to the tens column. (This accounts for the 2 above the 5 in the rewrite.) By adding the tens column, we get 2 + 5 + 8 + 6 + 0 = 21. We can place the 1 in the answer line and carry the 2 to the hundreds column. (This accounts for the 2 above the 4 in the above problem.) By adding the hundreds column, we get 2 + 4 + 5 + 7 + 6 = 24. We can place the 4 in the answer line and carry the 2 to the thousands column. (This accounts for the 2 above the 3 in the rewrite of the problem.) By adding the thousands column, we get 2 + 3 + 2 + 3 + 4 = 14. Since we have no more columns to add, we can write the 14 in our answer line. The final answer is 14,416.

Example 3

Simplify the following problems.

(a)   235 + 567

```
              1 1
  235         235
+ 567       + 567
              802
```

By adding the ones column, we get 5 + 7 = 12. We can place the 2 in our answer line and carry the 1 to the tens column. (This would account

for the 1 above the 3 in the example.) By adding the tens column, we get 1 + 3 + 6 = 10. We can place the 0 in the answer line and carry the 1 to the hundreds column. (This would account for the 1 above the 2 in the rewrite.) By adding the hundreds column, we get 1 + 2 + 5 = 8. The final answer is 802.

(b)   896 + 561 + 367

```
              2 1
  896         896
  561         561
+ 367       + 367
             1824
```

By adding the ones column, we get 6 + 1 + 7 = 14. We can place the 4 in our answer line and carry the 1 to the tens column. (This accounts for the 1 over the 9 in the rewrite.) By adding the tens column, we get 1 + 9 + 6 + 6 = 22. We can place the 2 in the answer line and carry the 2 to the hundreds column. (This accounts for the 2 above the 8 in the rewrite of the problem.) By adding the hundreds column, we get 2 + 8 + 5 + 3 = 18. Since we have no more columns to add, we can write the 18 in the answer line. The final answer is 1,824.

(c)   4,675 + 3,407 + 5,671

```
               1 1 1
  4675         4675
  3407         3407
+ 5671       + 5671
              13753
```

By adding the ones column, we get 5 + 7 + 1 = 13. We can place the 3 in the answer line and carry the 1 to the tens column. (This accounts for the 1 above the 7 in the rewrite.) By adding the tens column, we get 1 + 7 + 0 + 7 = 15. We can write the 5 in the answer line and carry the 1 to the hundreds column. (This accounts for the 1 above the 6 in the problem). By adding the hundreds column, we get 1 + 6 + 4 + 6 = 17. We can write the 7 in the answer line and carry the 1 to the thousands column. (This accounts for the 1 above the 4 in the rewrite.) By adding the thousands column, we get 1 + 4 + 3 + 5 = 13. Since we have no more columns to add, we can write the 13 in our answer line. The final answer is 13,753.

(d) 348 + 475 + 981 + 296

$$
\begin{array}{r}
348 \\
475 \\
981 \\
+\ 296 \\
\end{array}
\longrightarrow
\begin{array}{r}
{\scriptstyle 3\,2} \\
348 \\
475 \\
981 \\
+\ 296 \\
\hline
2100 \\
\end{array}
$$

By adding the ones column, we get 8 + 5 + 1 + 6 = 20. We can place the 0 in the answer line and carry the 2 to the tens column. (This accounts for the 2 above the 4 in this problem.) By adding the tens column, we get 2 + 4 + 7 + 8 + 9 = 30. We can place the 0 in the answer line and carry the 3 to the hundreds column. (This accounts for 3 above the 3 in the example.) By adding the hundreds column, we get 3 + 3 + 4 + 9 + 2 = 21. Since we have no more columns to add, we can write the 21 in our answer line. The final answer is 2,100.

Questions

Add the following numbers.

1. Add 346 and 521.

2. Add 2,345 and 5,432.

3. Add 346 and 419.

4. Add 6,793 and 2,406

5. Add 9,874 and 6,751.

6. 245 + 912 + 357

7. 4,376 + 1,994 + 8,953

8. 15,274 + 98,407

9. 4,275 + 7,136 + 3,983 + 5,648

10. 67,134 + 34,673 + 25,876

Answers

1. 867

$$
\begin{array}{r}
346 \\
+\ 521 \\
\hline
867 \\
\end{array}
$$

2. 7,777

$$
\begin{array}{r}
2345 \\
+\ 5432 \\
\hline
7777 \\
\end{array}
$$

3. 765

$$
\begin{array}{r}
{\scriptstyle 1} \\
346 \\
+\ 419 \\
\hline
765 \\
\end{array}
$$

4. 9,199

    1
    6793
    + 2406
    ─────
    9199

5. 16,625

    1 1
    9874
    + 6751
    ─────
    16625

6. 1,514

    1 1
    245
    912
    + 357
    ─────
    1514

7. 15,323

    2 21
    4376
    1994
    + 8953
    ─────
    15,323

8. 113,681

    1   1
    15274
    + 98407
    ─────
    113681

9. 21,042

    2 22
    4275
    7136
    3983
    + 5648
    ─────
    21042

10. 127,683

    1 1 1
    67134
    34673
    + 25876
    ─────
    127683

SUBTRACTION OF WHOLE NUMBERS

Now that we are "experts" at adding numbers, we want to discuss the other operations that can be used with whole numbers. In this section, we will consider subtraction. As with addition, in subtraction we will write our numbers in column format, with the place values lined up. We will subtract the bottom number from the top number (if possible), moving right to left. If the subtraction is not possible, we can borrow one from the next larger column. In the following examples, we will borrow where necessary.

Example 1

Subtract the following pairs of numbers.

(a) 786 – 135

$$
\begin{array}{r} 786 \\ -\ 135 \\ \hline \end{array}
\longrightarrow
\begin{array}{r} 786 \\ -\ 135 \\ \hline 651 \end{array}
$$

By subtracting the ones column, we get $6 - 5 = 1$. We can write this value on our answer line. By subtracting the tens column, we get $8 - 3 = 5$. We can write this value on our answer line. By subtracting the hundreds column, we get $7 - 1 = 6$. We can write this value on our answer line. The final answer is 651. Notice that we do not need to borrow in this example.

(b) $8,967 - 1,364$

$$
\begin{array}{r}
8967 \\
- 1364 \\
\end{array}
\longrightarrow
\begin{array}{r}
8967 \\
- 1364 \\
\hline
7603 \\
\end{array}
$$

By subtracting the ones column, we get $7 - 4 = 3$. We can write this value on our answer line. By subtracting the tens column, we get $6 - 6 = 0$. We can write this value on our answer line. By subtracting the hundreds column, we get $9 - 3 = 6$. We can write this value on our answer line. By subtracting the thousands column, we get $8 - 1 = 7$. We can write this value on our answer line. Our final answer is 7,603. Again, in this example, we did not need to borrow.

(c) $4,572 - 3,469$

$$
\begin{array}{r}
4572 \\
- 3469 \\
\end{array}
\longrightarrow
\begin{array}{r}
1 \\
4562 \\
- 3469 \\
\hline
1103 \\
\end{array}
$$

If we try to subtract the ones column, we see that we have a difficulty ($2 - 9$ cannot be done using whole numbers). In order to do this subtraction, we need to borrow 1 from the tens column. This would change the 7 to a 6 in the tens position and change 2 into 12 in the ones position. We are now able to subtract the ones column. This will give $12 - 9 = 3$. We can write this value on our answer line. By subtracting the tens column, we get $6 - 6 = 0$. We can write this value on our answer line. By subtracting the hundreds column, we get $5 - 4 = 1$. We can write this value on our answer line. By subtracting the thousands column, we get $4 - 3 = 1$. We can write this value on our answer line. The final value is 1,103.

(d) $8,737 - 4,364$

$$
\begin{array}{r}
8737 \\
- 4364 \\
\end{array}
\longrightarrow
\begin{array}{r}
1 \\
8637 \\
- 4364 \\
\hline
4373 \\
\end{array}
$$

By subtracting the ones column, we get $7 - 4 = 3$. We can write this value on our answer line. In order to subtract the tens column, we need to borrow 1 from the hundreds column. The 7 becomes a 6 and the 3 becomes a 13. Therefore, subtracting the tens column gives $13 - 6 = 7$. This value can be placed on our answer line. By subtracting the hundreds column, we get $6 - 3 = 3$. We can write this value on our answer line. By subtracting the thousands column, we get $8 - 4 = 4$. We can write this value on our answer line. The final answer is 4,373.

Example 2

Subtract the following numbers. Borrow when necessary.

(a) $3,574 - 1,647$

$$
\begin{array}{r}
3574 \\
- 1647 \\
\end{array}
\longrightarrow
\begin{array}{r}
1\ 1 \\
2564 \\
- 1647 \\
\hline
1927 \\
\end{array}
$$

In order to subtract the ones column, we need to borrow 1 from the tens column. This causes the 7 to become a 6 and the 4 to become 14. Subtracting the ones column gives $14 - 7 = 7$. We can write this value on our answer line. By subtracting the tens column, we get $6 - 4 = 2$. We can write this value on our answer line. In order to subtract the hundreds column, we need to borrow from the thousands column. This causes the 3 to become a 2 and the 5 to become a 15.

Subtracting the hundreds column gives $15 - 6 = 9$. We can write this value on our answer line. By subtracting the thousands column, we get $2 - 1 = 1$. We can write this value on our answer line. The final answer is 1,927.

(b) $9,836 - 3,256$

$$
\begin{array}{r}
9836 \\
- \ 3256 \\
\end{array}
\quad \longrightarrow \quad
\begin{array}{r}
\overset{1}{} \\
9736 \\
- \ 3256 \\
\hline
6580 \\
\end{array}
$$

By subtracting the ones column, we get $6 - 6 = 0$. We can write this value on our answer line. In order to subtract the tens column, we need to borrow from the hundreds column. This would cause the 8 to become a 7 and the 3 to become a 13. Therefore, we get $13 - 5 = 8$. We can write this value on our answer line. By subtracting the hundreds column, we get $7 - 2 = 5$. We can write this value on our answer line. By subtracting the thousands column, we get $9 - 3 = 6$. We can write this value on our answer line. Our final answer is 6,580.

(c) $857 - 289$

$$
\begin{array}{r}
857 \\
- \ 289 \\
\end{array}
\quad \longrightarrow \quad
\begin{array}{r}
\overset{1\,1}{} \\
747 \\
- \ 289 \\
\hline
568 \\
\end{array}
$$

In order to subtract the ones column, we need to borrow 1 from the tens column. This causes the 5 to become a 4 and the 7 to become 17. Therefore, we will get $17 - 9 = 8$. We can write this value on our answer line. In order to subtract the tens column, we need to borrow 1 from the hundreds column. This would cause the 8 to become a 7 and the 4 to become

a 14. Therefore, we will get $14 - 8 = 6$. We can write this value on our answer line. By subtracting the hundreds column, we get $7 - 2 = 5$. We can write this value on our answer line. The final answer is 568.

(d) $6,758 - 5,964$

$$
\begin{array}{r}
6758 \\
- \ 5964 \\
\end{array}
\quad \longrightarrow \quad
\begin{array}{r}
\overset{1\,1}{} \\
5658 \\
- \ 5964 \\
\hline
794 \\
\end{array}
$$

By subtracting the ones column, we get $8 - 4 = 4$. We can write this value on our answer line. In order to subtract the tens column, we need to borrow 1 from the hundreds column. This causes the 7 to become a 6 and the 5 becomes a 15. Therefore, we get $15 - 6 = 9$. We can write this value on our answer line. In order to subtract the hundreds column, we need to borrow 1 from the thousands column. This causes the 6 to become a 5 and the 6 to become a 16. Therefore, we get $16 - 9 = 7$. We can write this value on our answer line. By subtracting the thousands column, we get $5 - 5 = 0$. Though writing 0 is not necessary, we can write this value on our answer line. The final answer is 794.

Example 3

Simplify the following problems.

(a) $8,742 - 3,467$

$$
\begin{array}{r}
8742 \\
- \ 3467 \\
\end{array}
\longrightarrow
\begin{array}{r}
8742 \\
- \ 3467 \\
\end{array}
\longrightarrow
$$

$$
\longrightarrow
\begin{array}{r}
\overset{1\,1}{} \\
8632 \\
- \ 3467 \\
\hline
5275 \\
\end{array}
$$

In order to subtract the ones column, we need to borrow 1 from the tens column. This causes the 4 to become a 3 and the 2 to become a 12. Therefore, we get $12 - 7 = 5$. We can write this value on our answer line. In order to subtract the tens column, we need to borrow 1 from the hundreds column. This causes the 7 to become a 6 and the 3 to become 13. Therefore, we get $13 - 6 = 7$. We can write this value on our answer line. By subtracting the hundreds column, we get $6 - 4 = 2$. We can write this value on our answer line. By subtracting the thousands column, we get $8 - 3 = 5$. We can write this value on our answer line. Our final answer is 5,275.

(b) $974 - 734$

$$
\begin{array}{r}
974 \\
- 734 \\
\end{array}
\longrightarrow
\begin{array}{r}
974 \\
- 734 \\
\hline
240 \\
\end{array}
$$

By subtracting the ones column, we get $4 - 4 = 0$. We can write this value on our answer line. By subtracting the tens column, we get $7 - 3 = 4$. We can write this value on our answer line. By subtracting the hundreds column, we get $9 - 7 = 2$. We can write this value on our answer line. Our final answer is 240.

(c) $7,645 - 6,713$

$$
\begin{array}{r}
7645 \\
- 6713 \\
\end{array}
\longrightarrow
\begin{array}{r}
{\scriptstyle 1} \\
6645 \\
- 6713 \\
\hline
932 \\
\end{array}
$$

By subtracting the ones column, we get $5 - 3 = 2$. This value can be placed on our answer line. By subtracting the tens column, we get $4 - 1 = 3$. This value can be placed on our answer line. In order to subtract the hundreds column, we need to borrow 1 from the thousands column. Therefore, the 7 becomes a 6 and the 6 becomes a 16. We now get $16 - 7 = 9$. We can write this value on our answer line. By subtracting the thousands column, we get $6 - 6 = 0$. Our final answer is 932.

(d) $945 - 219$

$$
\begin{array}{r}
945 \\
- 219 \\
\end{array}
\longrightarrow
\begin{array}{r}
{\scriptstyle 1} \\
935 \\
- 219 \\
\hline
726 \\
\end{array}
$$

In order to subtract the ones column, we need to borrow 1 from the tens column. Therefore, the 4 becomes a 3 and the 5 becomes a 15. We now get $15 - 9 = 6$. We can write this value on our answer line. By subtracting our tens column, we get $3 - 1 = 2$. We can write this value on our answer line. By subtracting the hundreds column, we get $9 - 2 = 7$. We can write this value on our answer line. Our final answer is 726.

Questions

Subtract the following pairs of numbers.

1. $908 - 603$

2. $9,846 - 1,345$

3. $4,256 - 2,133$

4. $3,457 - 1,246$

5. $89,725 - 68,413$

6. $963 - 249$

7. $2,795 - 1,257$

8. 5,637 – 4,258

9. 874 – 793

10. 85,267 – 25,456

Answers

1. 305

   908
   – 603
   ─────
   305

2. 8,501

   9846
   – 1345
   ─────
   8501

3. 2,123

   4256
   – 2133
   ─────
   2123

4. 2,211

   3457
   – 1246
   ─────
   2211

5. 21,312

   89725
   – 68413
   ──────
   21312

6. 714

   5 1
   963
   – 249
   ─────
   714

7. 1,538

   8 1
   2795
   – 1257
   ─────
   1538

8. 1,379

   3 12 1
   5637
   – 4258
   ─────
   1379

9. 81

   7 16 1
   874
   – 793
   ─────
   81

10. 59,811

   7 14 1
   85267
   – 25456
   ──────
   59811

MULTIPLICATION OF WHOLE NUMBERS

The next operation that we want to discuss is multiplication. As with addition and subtraction, we will write our problems in a column format. We will begin by multiplying the ones digit in the second number by each digit in the first number. Then, if it exists, we will repeat this process with the tens digit in the second number. This action will continue until we have used each digit in the second number. Remember to carry where necessary. After multiplying the next digits, we need to add in anything carried over from the preceding product. If the second number contains more than one digit, our final step is to add down each column to get our final answer. In the following example, we will address a variety of problems. In this book, we will use an ($\times$) to denote multiplication.

Example 1

Multiply the following pairs of numbers.

(a) 376×2

$$
\begin{array}{r}
376 \\
\times\ 2 \\
\hline
\end{array}
\longrightarrow
\begin{array}{r}
{\scriptstyle 1\,1} \\
376 \\
\times\ 2 \\
\hline
752
\end{array}
$$

We begin by multiplying the 2 and the 6 together, producing 12. We write the 2 on our answer line and carry the 1. (This accounts for the 1 appearing above the 7 in this problem.) We then multiply the 2 and the 7 together, getting 14. Since we carried 1 from the preceding product, this value must be added to the 14. This will give 15. We can place the 5 on our answer line and carry the 1. (This accounts for the 1 appearing above the 3 in this ex-

ample.) This process is continued by multiplying the 2 and the 3 together. This results in 6. When we add the 1 that was carried over, our result for this step would be 7. We can write this value on our answer line. Since the second number only contains one digit and we have multiplied this digit throughout the first number, we know that our problem is complete. Our overall answer is 752.

(b) $1,205 \times 7$

$$
\begin{array}{r}
1205 \\
\times\ 7 \\
\hline
\end{array}
\longrightarrow
\begin{array}{r}
{\scriptstyle 1\ 3} \\
1205 \\
\times\ 7 \\
\hline
8435
\end{array}
$$

We begin by multiplying the 7 and the 5 together, producing 35. We write the 5 on our answer line and carry the 3. (This accounts for the 3 appearing above the 0 in this problem.) We then multiply the 7 and the 0 together, getting 0. Since we carried 3 from the preceding product, this value must be added to the 0. This gives 3. We can place the 3 on our answer line. This process is continued by multiplying the 7 and the 2 together. This results in 14. We can write the 4 on our answer line and carry the 1. (This accounts for the 1 appearing above the 1 in this example.) Next we must multiply the 7 and the 1 together. This gives 7. When we add the 1 that was carried, we get an 8 for a result in this step. Since the second number only contains one digit and we have multiplied this digit throughout the first number, we know that our problem is complete. Our overall answer is 8,435.

(c) 679×34

$$
\begin{array}{r}
679 \\
\times\ \ 34 \\
\hline
\end{array}
\longrightarrow
\begin{array}{r}
{}^{3\ 3} \\
679 \\
\times\ \ 34 \\
\hline
2716 \\
\end{array}
\longrightarrow
$$

$$
\longrightarrow
\begin{array}{r}
{}^{2\ 2} \\
679 \\
\times\ \ 34 \\
\hline
2716 \\
+\ 20370 \\
\hline
\end{array}
\longrightarrow
\begin{array}{r}
{}^{1} \\
2716 \\
+\ 20370 \\
\hline
23086 \\
\end{array}
$$

We begin by multiplying the 4 and the 9 together, producing 36. We write the 6 on our answer line and carry the 3. (This accounts for the 3 appearing above the 7 in the second rewrite of this problem.) We then multiply the 4 and the 7 together, getting 28. We now need to add the 3 that was carried over to this result. This gives 31. We can place the 1 on our "answer line" and carry the 3. This process is continued by multiplying the 4 and the 6 together. This results in 24 and when we add in the 3 that was carried over, we get 27. We can write this value on our "answer line." So far in this problem we have placed quotes around the phrase "answer line." The purpose of this is to denote that we are talking about the "answer" of this part (or step) in our problem. Since the second number contains two digits, we must multiply each digit by the first number. At this point we have only multiplied the ones digit in the second number by the first number. Now we have to do the same thing with the tens digit of the second number. In order to denote that we are multiplying by the tens digit, we have placed a 0 in the ones digit as a place holder on the "answer line" of the

second part of this example. (This is the italicized zero (*0*) appearing in the third rewrite of this example.) We then take 3 and multiply it by 9. This results in 27. We can place the 7 in our "answer line" for this stage of the problem and carry the 2. Next, we multiply the 3 and the 7. In this case, we get 21. When we add in the 2 that was carried over, we get 23. We can place the 3 on our "answer line" and carry the 2. Now we can multiply the 3 and the 6 to get 18. When we add in the 2 that was carried over, we get 20. We can write this value on our "answer line." As we discussed above, the final stage of this problem is to add the two "answer lines" together. Keep in mind that we also need to carry here when necessary. Our overall answer is 23,086.

(d) $3,451 \times 67$

$$
\begin{array}{r}
3451 \\
\times\ \ 67 \\
\hline
\end{array}
\longrightarrow
\begin{array}{r}
{}^{3\ 3} \\
3451 \\
\times\ \ 67 \\
\hline
24157 \\
\end{array}
\longrightarrow
$$

$$
\longrightarrow
\begin{array}{r}
{}^{2\ 3} \\
3451 \\
\times\ \ 67 \\
\hline
24157 \\
+\ 207060 \\
\hline
\end{array}
\longrightarrow
\begin{array}{r}
{}^{1\ 1} \\
24157 \\
+\ 207060 \\
\hline
231217 \\
\end{array}
$$

We begin by multiplying the 7 and the 1 together, producing 7. We write this value on our "answer line." We then multiply the 7 and the 5 together, getting 35. We can place the 5 on our "answer line" and carry the 3. (This accounts for the 3 above the 4 in the second rewrite of this problem.) We can continue this process by multiplying the 7 and the 4 together. This results in 28. When we add in the 3 that was carried over, we get 31. We can write the 1 on our "answer line" and carry the 3. (This accounts for the 3 appearing above the 3 in the second copy of the example.) By multiplying the 7 and the 3 together, we get 21. When we add in the 3 that was carried over, we get a result of 24. We can write this value on our "answer line." Again, in this example, we have used quotes around the phrase "answer line" to denote that we are talking about the "answer" of this part (or step) in our problem. Since the second number contains two digits, we must multiply each digit by the first number. At this point, we have only multiplied the ones digit in the second number by the first number. Now we have to do the same thing with the tens digit of the second number. Again, we have placed a 0 in the ones digit on the "answer line" for the second step of this problem. (This is the italicized zero (*0*) appearing in the third rewrite of this example.) We then take 6 and multiply it by 1. This results in 6, and we can place this value on our "answer line." Next, we multiply the 6 and the 5. In this case we get 30. We can place the 0 on our "answer line" and carry the 3. (This accounts for the 3 appearing above the 4 in the third rewrite of the above problem.) Now we can multiply the 6 and the 4 to get 24. When we add in the 3 that was carried over, we get 27. We can write the 7 on our "answer line" and carry the 2. (This accounts for the 2 appearing above the 3 in the third rewrite of this problem.) When we multiply the 6 and the 3 together, we get 18. When we add in the 2 that was carried over, we get 20. We can place this value on our

"answer line." The final stage of this problem is to add the two "answer lines" together. Keep in mind that we also need to carry here when necessary. Our final answer is 231,217.

Example 2

Multiply the following numbers.

(a) 304×25

$$
\begin{array}{r} 304 \\ \times\ 25 \\ \hline \end{array}
\longrightarrow
\begin{array}{r} {}^{2} \\ 304 \\ \times\ 25 \\ \hline 1520 \end{array}
\longrightarrow
$$

$$
\longrightarrow
\begin{array}{r} 304 \\ \times\ 25 \\ \hline 1520 \\ +\ 608\mathit{0} \\ \hline \end{array}
\longrightarrow
\begin{array}{r} {}^{1} \\ 1520 \\ +\ 6080 \\ \hline 7600 \end{array}
$$

We begin by multiplying the ones digit in the second number by the first number. In this problem, we would get $304 \times 5 = 1,520$. We place this value on our first "answer line." Remember to carry when necessary. We then need to take the tens digit in the second number and multiply it by the first number. In this example we would get $304 \times 2 = 608$. We have placed a zero (*0*) in the ones position of our second "answer line" in order to keep the proper columns. We also did this in the previous examples. Therefore, our second "answer line" would be 6,08*0*. To finish up this problem we must add the two "answer lines" together. If we again remember to carry where needed, we will get a final answer of 7600. Notice that in the work above we have shown all carried numbers in italics.

(b) 59×67

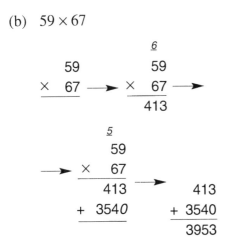

$$
\begin{array}{r}
59 \\
\times\ 67 \\
\hline
413 \\
+\ 3540 \\
\hline
3953
\end{array}
$$

We begin by multiplying the ones digit in the second number by the first number. In this problem, we would get $59 \times 7 = 413$. We place this value on our first "answer line." Remember to carry when necessary. We then need to take the tens digit in the second number and multiply it by the first number. In this example, we would get $59 \times 6 = 354$. We have placed a zero (*0*) in the ones position of our second "answer line" in order to keep the proper columns. Therefore, our second "answer line" would be 3,54*0*. To finish up this problem, we must add the two "answer lines" together. If we again remember to carry where needed, we will get a final answer of 3,953. Notice that in the work above, we have shown all carried numbers in italics.

(c) 382×19

$$
\begin{array}{r}
{\it 71} \\
382 \\
\times\ 19 \\
\hline
3438
\end{array}
$$

$$
\begin{array}{r}
382 \\
\times\ 19 \\
\hline
3438 \\
+\ 3820 \\
\hline
\end{array}
\qquad
\begin{array}{r}
{\it 1} \\
3438 \\
+\ 3820 \\
\hline
7258
\end{array}
$$

We begin by multiplying the ones digit in the second number by the first number. In this problem, we would get $382 \times 9 = 3,438$. We place this value on our first "answer line." Remember to carry when necessary. We then need to take the tens digit in the second number and multiply it by the first number. In this example, we would get $382 \times 1 = 382$. We have placed a zero (*0*) in the ones position of our second "answer line" in order to keep the proper columns. Therefore, our second "answer line" would be 3,82*0*. To finish up this problem, we must add the two "answer lines" together. If we again remember to carry where needed, we will get a final answer of 7,258. Notice that in the work above we have shown all carried numbers in italics.

(d) 736×142

$$
\begin{array}{r}
736 \\
\times\ 142 \\
\hline
\end{array}
\qquad
\begin{array}{r}
{\it 1} \\
736 \\
\times\ 142 \\
\hline
1472
\end{array}
\qquad
\begin{array}{r}
736 \\
\times\ 142 \\
\hline
1472 \\
+\ 29440 \\
\hline
\end{array}
$$

$$
\begin{array}{r}
736 \\
\times\ 142 \\
\hline
1472 \\
29440 \\
+\ 73600 \\
\hline
\end{array}
\qquad
\begin{array}{r}
{\it 111} \\
1472 \\
29440 \\
+\ 73600 \\
\hline
104512
\end{array}
$$

We begin by multiplying the ones digit in the second number by the first number. In this problem, we would get $736 \times 2 = 1,472$. We place this value on our first "answer line." Remember to carry when necessary. We then need to take the tens digit in the second number and multiply it by the first number. In this example, we would get $736 \times 4 = 2,944$. We have placed a zero (0) in the ones position of our second "answer line" in order to keep the proper columns. Therefore, our second "answer line" would be 29,44*0*. We now need to take the hundreds digit in the second number and multiply it by the first number. In this example, we would get 736 × 1 = 736. We have placed zeros (0) in the ones and tens positions of our third "answer line" in order to keep the proper columns. Therefore, our third "answer line" would be 73,6*00*. To finish up this problem, we must add the three "answer lines" together. If we again remember to carry where needed, we will get a final answer of 104,512. Notice that in the work above we have shown all carried numbers in italics.

Example 3

Simplify the following problems.

(a) 89×61

$$\begin{array}{r} 89 \\ \times \ 61 \\ \hline \end{array} \longrightarrow \begin{array}{r} 89 \\ \times \ 61 \\ \hline 89 \end{array} \longrightarrow$$

$$\begin{array}{r} \textit{5} \\ 89 \\ \times \ 61 \\ \hline 89 \\ + \ 534\textit{0} \\ \hline \end{array} \longrightarrow \begin{array}{r} \textit{1} \\ 89 \\ + \ 5340 \\ \hline 5429 \end{array}$$

We begin by multiplying the ones digit in the second number by the first number. In this problem, we would get 89 × 1 = 89. We place this value on our first "answer line." Remember to carry when necessary. We then need to take the tens digit in the second number and multiply it by the first number. In this example, we

would get 89 × 6 = 534. We have placed a zero (0) in the ones position of our second "answer line" in order to keep the proper columns. Therefore, our second "answer line" would be 5,34*0*. To finish up this problem, we must add the two "answer lines" together. If we again remember to carry where needed, we will get a final answer of 5,429. Notice that in the work above we have shown all carried numbers in italics.

(b) 715×36

$$\begin{array}{r} 715 \\ \times \ 36 \\ \hline \end{array} \longrightarrow \begin{array}{r} \textit{3} \\ 715 \\ \times \ 36 \\ \hline 4290 \end{array} \longrightarrow$$

$$\begin{array}{r} \textit{1} \\ 715 \\ \times \ 36 \\ \hline 4290 \\ + \ 2145\textit{0} \\ \hline \end{array} \longrightarrow \begin{array}{r} \textit{1} \\ 4290 \\ + \ 21450 \\ \hline 25740 \end{array}$$

We begin by multiplying the ones digit in the second number by the first number. In this problem, we would get 715 × 6 = 4,290. We place this value on our first "answer line." Remember to carry when necessary. We then need to take the tens digit in the second number and multiply it by the first number. In this example, we would get 715 × 3 = 2,145. We have placed a zero (0) in the ones position of our second "answer line" in order to keep the proper columns. Therefore, our second "answer line" would be 21,45*0*. To finish up this problem, we must add the two "answer lines" together. If we again remember to carry where needed, we will get a final answer of 25,740. Notice that in the work above we have shown all carried numbers in italics.

(c) 123×45

$$
\begin{array}{r}
123 \\
\times \ 45 \\
\end{array}
\longrightarrow
\begin{array}{r}
\textit{1 1} \\
123 \\
\times \ 45 \\
\hline
615 \\
\end{array}
\longrightarrow
$$

$$
\longrightarrow
\begin{array}{r}
\textit{1} \\
123 \\
\times \ 45 \\
\hline
615 \\
+ \ 492\textit{0} \\
\end{array}
\longrightarrow
\begin{array}{r}
\textit{1} \\
615 \\
+ \ 4920 \\
\hline
5535 \\
\end{array}
$$

We begin by multiplying the ones digit in the second number by the first number. In this problem, we would get $123 \times 5 = 615$. We place this value on our first "answer line." Remember to carry when necessary. We then need to take the tens digit in the second number and multiply it by the first number. In this example, we would get $123 \times 4 = 492$. We have placed a zero ($\textit{0}$) in the ones position of our second "answer line" in order to keep the proper columns. Therefore, our second "answer line" would be 4,92$\textit{0}$. To finish up this problem, we must add the two "answer lines" together. If we again remember to carry where needed, we will get a final answer of 5,535. Notice that in the work above we have shown all carried numbers in italics.

(d) 632×113

$$
\begin{array}{r}
632 \\
\times \ 113 \\
\end{array}
\longrightarrow
\begin{array}{r}
632 \\
\times \ 113 \\
\hline
1896 \\
\end{array}
\longrightarrow
\begin{array}{r}
632 \\
\times \ 113 \\
\hline
1896 \\
+ \ 632\textit{0} \\
\end{array}
$$

$$
\longrightarrow
\begin{array}{r}
632 \\
\times \ 113 \\
\hline
1896 \\
6320 \\
+ \ 632\textit{00} \\
\end{array}
\longrightarrow
\begin{array}{r}
\textit{1 1} \\
1896 \\
6320 \\
+ \ 63200 \\
\hline
71416 \\
\end{array}
$$

We begin by multiplying the ones digit in the second number by the first number. In this problem, we would get $632 \times 3 = 1,896$. We place this value on our first "answer line." Remember to carry when necessary. We then need to take the tens digit in the second number and multiply it by the first number. In this example, we would get $632 \times 1 = 632$. We have placed a zero ($\textit{0}$) in the ones position of our second "answer line" in order to keep the proper columns. Therefore, our second "answer line" would be 6,32$\textit{0}$. We now need to take the hundreds digit in the second number and multiply it by the first number. In this example, we would get $632 \times 1 = 632$. We have placed zeroes ($\textit{0}$) in the ones and tens positions of our third "answer line" in order to keep the proper columns. Therefore, our third "answer line" would be 63,2$\textit{00}$. To finish up this problem, we must add the three "answer lines" together. If we again remember to carry where needed, we will get a final answer of 71,416. Notice that in the work above we have shown all carried numbers in italics.

Questions

Multiply the following pairs of numbers.

1. 543×9

2. $6,287 \times 4$

3. $1,247 \times 3$

4. 496×12

5. $8,157 \times 46$

6. 370×51

7. 607×32

8. 816×79

9. 434×65

10. 963×246

Answers

1. 4,887

   $$
   \begin{array}{r}
   32 \\
   543 \\
   \times \quad 9 \\
   \hline
   4887 \\
   \end{array}
   $$

2. 25,148

   $$
   \begin{array}{r}
   13\ 2 \\
   6287 \\
   \times \quad 4 \\
   \hline
   25148 \\
   \end{array}
   $$

3. 3,741

   $$
   \begin{array}{r}
   1\ 2 \\
   1247 \\
   \times \quad 3 \\
   \hline
   3741 \\
   \end{array}
   $$

4. 5,952

   $$
   \begin{array}{r}
   11 \\
   496 \\
   \times \quad 12 \\
   \hline
   992 \\
   +\ 496 \\
   \hline
   5952 \\
   \end{array}
   $$

5. 375,222

   $$
   \begin{array}{r}
   8157 \\
   \times \quad 46 \\
   \hline
   48942 \\
   +\ 32628 \\
   \hline
   375222 \\
   \end{array}
   $$

6. 18,870

   $$
   \begin{array}{r}
   370 \\
   \times\ 51 \\
   \hline
   370 \\
   +\ 1850 \\
   \hline
   18870 \\
   \end{array}
   $$

7. 19,424

   $$
   \begin{array}{r}
   607 \\
   \times\ 32 \\
   \hline
   1214 \\
   +\ 1821 \\
   \hline
   19424 \\
   \end{array}
   $$

8. 64,464

   $$
   \begin{array}{r}
   816 \\
   \times\ 79 \\
   \hline
   7344 \\
   +\ 5712 \\
   \hline
   64464 \\
   \end{array}
   $$

9. 28,210

      434

      × 65

      2170

    + 2604

      28210

10. 236,898

        963

      × 246

      5778

      3852

    + 1926

      236898

DIVISION OF WHOLE NUMBERS

The last operation that we want to cover is division. By writing the problems in long division format, we can compare digits. Here we are looking to see how many times a number goes into another. That is, what must a number be multiplied by to get the other? Hopefully this can be an exact number. If not, we have a **remainder**. In this book, we will write "R" and then the value to denote a remainder. Here we want the largest value possible without exceeding the number. This is a quick over-

view of the method that we will use. In the examples below, we will refine this technique.

Example 1

Divide the following numbers.

(a) 638 / 2

$$2\overline{)638} \rightarrow 2\overline{)638}\;^{3} \rightarrow$$
$$\underline{-6}$$
$$3$$

$$\rightarrow 2\overline{)638}\;^{31} \rightarrow 2\overline{)638}\;^{319}$$
$$\underline{-6} \qquad\qquad \underline{-6}$$
$$3 \qquad\qquad 3$$
$$\underline{-2} \qquad\qquad \underline{-2}$$
$$18 \qquad\qquad 18$$
$$\qquad\qquad \underline{-18}$$
$$\qquad\qquad 0$$

In this example we are dividing a number (638) by a one-digit number (2). Since we have written the problem in long-division format, we can compare the digits, moving from the left to the right one digit at a time. We want to consider the first digit in 638. How many times will 2 go into 6? That is, what must be multiplied by 2 in order to get 6? Hopefully, this can be exact. If not, we want to use the largest value that is possible without exceeding our target value. In this case we get 3. When we multiply 2 and 3 together we get 6. By writing this value under the first number and subtracting the column we get 6 − 6 = 0. We now bring down the next digit and start this process again. How many times will 2 go into 3? That is, what must be multiplied by 2 to get 3? In this case, we cannot find an exact value. The largest number possible (without exceeding a product of

3) would be 1. Two would be too large since $2 \times 2 = 4$ and 4 is greater than 3. Since $2 \times 1 = 2$, we can write this value under the first number and subtract the column. Here we get $3 - 2 = 1$. When we bring down the next digit (8), we get 18. We now want to consider how many times 2 goes into 18. That is, what must be multiplied by 2 to get 18? In this case, we get 9. If we multiply the 2 and the 9 together, we get 18. We can write this value under the first 18 and subtract. Here we will get $18 - 18 = 0$. In this example, we do not get a remainder. If needed (or wished), we can check our answer by multiplying that answer by the number to the left of the division bar. If our work is correct, we will get the number appearing under the division bar. In this problem, $319 \times 2 = 638$.

(b)   753 / 3

$$
3\overline{)753} \longrightarrow
\begin{array}{r}
2 \\
3\overline{)753} \\
-6 \\
\hline
15
\end{array}
\longrightarrow
$$

$$
\longrightarrow
\begin{array}{r}
25 \\
3\overline{)753} \\
-6 \\
\hline
15 \\
-15 \\
\hline
3
\end{array}
\longrightarrow
\begin{array}{r}
251 \\
3\overline{)753} \\
-6 \\
\hline
15 \\
-15 \\
\hline
3 \\
-3 \\
\hline
0
\end{array}
$$

Now we are dividing 753 by 3. Since we have written the problem in long-division format, we can compare the digits, moving from the left to the right one digit at a time. We want to consider the first digit. How many times will 3 go into 7? That is, what must be multiplied by 3 in order to

get 7? Hopefully, this can be exact. If not, we want to use the largest value that is possible without exceeding our target value. In this case, we get 2. When we multiply 3 and 2 together, we get 6. By writing this value under the first number and subtracting the column, we get $7 - 6 = 1$. We now bring down the next digit (5) and start this process again with 15. How many times will 3 go into 15? That is, what must be multiplied by 3 to get 15? In this case, we get 5. Since $3 \times 5 = 15$, we can write this value under the first number and subtract the column. Here we get $15 - 15 = 0$. We now bring down the next digit (3). Consider how many times 3 goes into 3. That is, what must be multiplied by 3 to get 3? In this case, we get 1. If we multiply the 3 and the 1 together, we get 3. We can write this value under the first number and subtract. Here we will get $3 - 3 = 0$. In this example, we do not get a remainder. We could check our answer by multiplying the answer by the number to the left of the division bar. If our work is correct, we will get the number appearing under the division bar. In this problem, $251 \times 3 = 753$.

(c)   475 / 5

$$
5\overline{)475} \longrightarrow
\begin{array}{r}
9 \\
5\overline{)475} \\
-45 \\
\hline
25
\end{array}
\longrightarrow
\begin{array}{r}
95 \\
5\overline{)475} \\
-45 \\
\hline
25 \\
-25 \\
\hline
0
\end{array}
$$

In this example, when we compare the 5 and the first digit under the division bar, we see that 5 cannot go into 4. It is impossible to multiply 5 by a whole number to get 4. This being the case, we now compare 5 and the first *two* digits in the number (47). How many times will 5 go into 47? That is, what must be multiplied by 5 to get 47? Here we get 9, because 9 $\times 5 = 45$. If we place this value under the first number and subtract, we get $47 - 45 = 2$, and bring down the next digit (5). Since we are looking at the first two numbers, we will place the 9 above the second digit from the left. Now we repeat this process. How many times will 5 go into 25? That is, what must we multiply 5 by to get 25? Our answer to this step is 5, since $5 \times 5 = 25$. Now we write this value under the first number and subtract the column. This gives $25 - 25 = 0$. As with the first two examples in this section, we do not get a remainder here. Checking our answer shows that it is correct.

(d)   2,975 / 7

$$\begin{array}{r} 7\,\overline{)\,2975} \end{array} \longrightarrow \begin{array}{r} 4 \\ 7\,\overline{)\,2975} \\ \underline{-28} \\ 17 \end{array} \longrightarrow$$

$$\longrightarrow \begin{array}{r} 42 \\ 7\,\overline{)\,2975} \\ \underline{-28} \\ 17 \\ \underline{-14} \\ 35 \end{array} \longrightarrow \begin{array}{r} 425 \\ 7\,\overline{)\,2975} \\ \underline{-28} \\ 17 \\ \underline{-14} \\ 35 \\ \underline{-35} \\ 0 \end{array}$$

Again, in this example, the number does not go into the first digit under the division bar. By comparing 7 and the first two digits (going left to right) in the number, we want to know how many times 7 goes into 29. Our answer is 4. Since $4 \times 7 = 28$, we can place this value under the first num-

ber and subtract the columns. Here we will get $29 - 28 = 1$ and bring down the next digit (7). We now repeat this method and see how many times 7 goes into 17. Our answer here is 2. Since we know that $7 \times 2 = 14$, we can place this value under the first number and subtract the columns. Here we get $17 - 14 = 3$ and bring down the next digit (5). Again, we repeat this method. How many times does 7 go into 35? Since $7 \times 5 = 35$, we can place this value under the first number and subtract the columns. In this example, we do not get a remainder. The answer check also works.

Example 2

Divide the following numbers.

(a)   564 / 12

$$\begin{array}{r} 12\,\overline{)\,564} \end{array} \longrightarrow \begin{array}{r} 4 \\ 12\,\overline{)\,564} \\ \underline{-48} \\ 84 \end{array} \longrightarrow$$

$$\longrightarrow \begin{array}{r} 47 \\ 12\,\overline{)\,564} \\ \underline{-48} \\ 84 \\ \underline{-84} \\ 0 \end{array}$$

Since it is not possible for 12 to go into 5, we consider how many times 12 goes into 56. Our answer here is 4. By multiplying 12 and 4 together, we get 48. We can place this value under the first number and subtract. Here we get $56 - 48 = 8$ and bring down the next digit (4). We now consider how many times 12 goes into 84. This gives 7, since $12 \times 7 = 84$. If we place this value under the first number and subtract, we get 0. There is no remainder in this problem.

(b) 4,768 / 32

$$32 \overline{)4768} \longrightarrow \begin{array}{r} 1 \\ 32 \overline{)4768} \\ -32 \\ \hline 156 \end{array} \longrightarrow$$

$$\longrightarrow \begin{array}{r} 14 \\ 32 \overline{)4768} \\ -32 \\ \hline 156 \\ -128 \\ \hline 288 \end{array} \longrightarrow \begin{array}{r} 149 \\ 32 \overline{)4768} \\ -32 \\ \hline 156 \\ -128 \\ \hline 288 \\ -288 \\ \hline 0 \end{array}$$

Since it is not possible for 32 to go into 4, we consider how many times 32 goes into 47. Our answer here is 1. By multiplying 32 and 1 together, we get 32. We can place this value under the first number and subtract. Here we get 47 – 32 = 15 and bring down the next digit (6). We now consider how many times 32 goes into 156. This gives 4, since 32 × 4 equals 128. When we place this value under the first number and subtract, we get 28. We bring down the next digit (8) and repeat this procedure. We now consider how many times 32 goes into 288. This gives 9, since 32 × 9 = 288. If we place this value under the first number and subtract, we get 0. There is no remainder in this problem.

(c) 16,468 / 46

$$46 \overline{)16468} \longrightarrow \begin{array}{r} 3 \\ 46 \overline{)16468} \\ -138 \\ \hline 266 \end{array} \longrightarrow$$

$$\longrightarrow \begin{array}{r} 35 \\ 46 \overline{)16468} \\ -138 \\ \hline 266 \\ -230 \\ \hline 368 \end{array} \longrightarrow \begin{array}{r} 358 \\ 46 \overline{)16468} \\ -138 \\ \hline 266 \\ -230 \\ \hline 368 \\ -368 \\ \hline 0 \end{array}$$

Since it is not possible for 46 to go into 1 or 16, we consider how many times 46 goes into 164. Our answer here is 3. By multiplying 46 and 3 together, we get 138. We can place this value under the first number and subtract. Here we get 164 – 138 = 26 and bring down the next digit (6). We now consider how many times 46 goes into 266. This gives 5, since 46 × 5 = 230. When we place this value under the first number and subtract, we get 36. We bring down the next digit (8) and repeat this procedure. We now consider how many times 46 goes into 368. This gives 8, since 46 × 8 = 368. If we place this value under the first number and subtract, we get 0. There is no remainder in this problem.

(d)   15,799 / 61

$$61\overline{)15799} \longrightarrow 61\overline{)\begin{array}{l}2\\15799\\-122\\\hline 359\end{array}} \longrightarrow$$

$$\longrightarrow 61\overline{)\begin{array}{l}25\\15799\\-122\\\hline 359\\-305\\\hline 549\end{array}} \longrightarrow 61\overline{)\begin{array}{l}259\\15799\\-122\\\hline 359\\-305\\\hline 549\\-549\\\hline 0\end{array}}$$

Since it is not possible for 61 to go into 1 or 15, we consider how many times 61 goes into 157. Our answer here is 2. By multiplying 61 and 2 together, we get 122. We can place this value under the first number and subtract. Here we get $157 - 122 = 35$ and bring down the next digit (9). We now consider how many times 61 goes into 359. This gives 5, since 61 $\times$ 5 = 305. When we place this value under the first number and subtract, we get 54. We bring down the next digit (9) and repeat this procedure. We now consider how many times 61 goes into 549. This gives 9, since 61 $\times$ 9 = 549. If we place this value under the first number and subtract, we get 0. There is no remainder in this problem.

Example 3

Divide the following numbers.

(a)   749 / 4

$$4\overline{)749} \longrightarrow 4\overline{)\begin{array}{l}1\\749\\-4\\\hline 34\end{array}} \longrightarrow$$

$$\longrightarrow 4\overline{)\begin{array}{l}18\\749\\-4\\\hline 34\\-32\\\hline 29\end{array}} \longrightarrow 4\overline{)\begin{array}{l}187\ R\ 1\\749\\-4\\\hline 34\\-32\\\hline 29\\-28\\\hline 1\end{array}}$$

We begin by considering how many times 4 goes into 7. Our answer is 1. Since $4 \times 1 = 4$, we can place this value under the first number and subtract. After subtracting and bringing down the next digit, we get 34. We now consider how many times 4 goes into 34. Our answer here is 8. Since $4 \times 8 = 32$, we can place this value under the first number and subtract. After subtracting and bringing down the next digit, we get 29. We now consider how many times 4 goes into 29. Our answer is 7. Since $4 \times 7 = 28$, we can place this value under the first number and subtract. This subtraction result is 1. Since we have no more digits in the first number, this result (1) becomes our remainder. Our overall answer is 187 R 1.

(b) 9,625 / 21

$$21\overline{)9625} \rightarrow 21\overset{4}{\overline{)9625}} \rightarrow$$
$$\underline{-\,84}$$
$$122$$

$$\rightarrow 21\overset{45}{\overline{)9625}} \rightarrow 21\overset{458\ R\ 7}{\overline{)9625}}$$
$$\underline{-\,84} \qquad\qquad \underline{-\,84}$$
$$122 \qquad\qquad 122$$
$$\underline{-\,105} \qquad\quad \underline{-\,105}$$
$$175 \qquad\qquad 175$$
$$\qquad\qquad\qquad \underline{-\,168}$$
$$\qquad\qquad\qquad 7$$

We begin by considering how many times 21 goes into 96. Our answer is 4. Since $21 \times 4 = 84$, we can place this value under the first number and subtract. After subtracting and bringing down the next digit, we get 122. We now consider how many times 21 goes into 122. Our answer here is 5. Since $21 \times 5 = 105$, we can place this value under the first number and subtract. After subtracting and bringing down the next digit, we get 175. We now consider how many times 21 goes into 175. Our answer is 8. Since $21 \times 8 = 168$, we can place this value under the first number and subtract. This subtraction result is 7. Since we have no more digits in the first number, this result (7) becomes our remainder. Our overall answer is 458 R 7.

(c) 15,245 / 57

$$57\overline{)15245} \rightarrow 57\overset{2}{\overline{)15245}} \rightarrow$$
$$\underline{-\,114}$$
$$384$$

$$\rightarrow 57\overset{26}{\overline{)15245}} \qquad 57\overset{267\ R\ 26}{\overline{)15245}}$$
$$\underline{-\,114} \qquad\qquad \underline{-\,114}$$
$$384 \qquad\qquad 384$$
$$\underline{-\,342} \qquad\quad \underline{-\,342}$$
$$425 \qquad\qquad 425$$
$$\qquad\qquad\qquad \underline{-\,399}$$
$$\qquad\qquad\qquad 26$$

We begin by considering how many times 57 goes into 152. Our answer is 2. Since $57 \times 2 = 114$, we can place this value under the first number and subtract. After subtracting and bringing down the next digit, we get 384. We now consider how many times 57 goes into 384. Our answer here is 6. Since $57 \times 6 = 342$, we can place this value under the first number and subtract. After subtracting and bringing down the next digit, we get 425. We now consider how many times 57 goes into 425. Our answer is 7. Since $57 \times 7 = 399$, we can place this value under the first number and subtract. This subtraction result is 26. Since we have no more digits in the first number, this result (26) becomes our remainder. Our overall answer is 267 R 26.

(d)   451,107 / 243

$$243 \overline{)451107} \longrightarrow 243 \overline{)\begin{matrix} 1 \\ 451107 \end{matrix}} \longrightarrow$$
$$\begin{matrix} -243 \\ \hline 2081 \end{matrix}$$

$$\longrightarrow 243 \overline{)\begin{matrix} 18 \\ 451107 \end{matrix}} \longrightarrow 243 \overline{)\begin{matrix} 185 \\ 451107 \end{matrix}} \longrightarrow$$
$$\begin{matrix} -243 \\ \hline 2081 \\ -1944 \\ \hline 1370 \end{matrix} \qquad \begin{matrix} -243 \\ \hline 2081 \\ -1944 \\ \hline 1370 \\ -1215 \\ \hline 1557 \end{matrix}$$

$$\longrightarrow 243 \overline{)\begin{matrix} 1856 \; R \; 99 \\ 451107 \end{matrix}}$$
$$\begin{matrix} -243 \\ \hline 2081 \\ -1944 \\ \hline 1370 \\ -1215 \\ \hline 1557 \\ -1458 \\ \hline 99 \end{matrix}$$

We begin by considering how many times 243 goes into 451. Our answer is 1. Since 243 × 1 = 243, we can place this value under the first number and subtract. After subtracting and bringing down the next digit, we get 2,081. We now consider how many times 243 goes into 2,081. Our answer here is 8. Since 243 × 8 = 1,944, we can place this value under the first number and subtract. After subtracting and bringing down the next digit, we get 1,370. We now con-sider how many times 243 goes into 1,370. Our answer is 5. Since 243 × 5 = 1,215, we can place this value under the first number and subtract. After subtracting and bringing down the next digit, we get 1,557. We now consider how many times 243 goes into 1,557. Our answer here is 6. Since 243 × 6 = 1,458, we can place this value under the first number and subtract. This subtraction result is 99. Since we have no more digits in the first number, this result (99) becomes our remainder. Our overall answer is 1,856 R 99.

Questions

Divide the following numbers.

1.   758 / 2

2.   868 / 7

3.   7,077 / 3

4.   1,280 / 5

5.   46,935 / 9

6.   515 / 2

7.   7,912 / 17

8.   7,799 / 25

9.   53,920 / 43

10.   122,444 / 128

Answers

1.   379

$$2\overline{)758} \longrightarrow \begin{array}{r} 3 \\ 2\overline{)758} \\ -6 \\ \hline 15 \end{array} \longrightarrow$$

$$\longrightarrow \begin{array}{r} 37 \\ 2\overline{)758} \\ -6 \\ \hline 15 \\ -14 \\ \hline 18 \end{array} \longrightarrow \begin{array}{r} 379 \\ 2\overline{)758} \\ -6 \\ \hline 15 \\ -14 \\ \hline 18 \\ -18 \\ \hline 0 \end{array}$$

2.   124

$$7\overline{)868} \longrightarrow \begin{array}{r} 1 \\ 7\overline{)868} \\ -7 \\ \hline 16 \end{array} \longrightarrow$$

$$\longrightarrow \begin{array}{r} 12 \\ 7\overline{)868} \\ -7 \\ \hline 16 \\ -14 \\ \hline 28 \end{array} \longrightarrow \begin{array}{r} 124 \\ 7\overline{)868} \\ -7 \\ \hline 16 \\ -14 \\ \hline 28 \\ -28 \\ \hline 0 \end{array}$$

3.   2,359

$$\begin{array}{r} 2 \\ 3\overline{)7077} \\ -6 \\ \hline 10 \end{array} \longrightarrow \begin{array}{r} 23 \\ 3\overline{)7077} \\ -6 \\ \hline 10 \\ -9 \\ \hline 17 \end{array} \longrightarrow$$

$$\longrightarrow \begin{array}{r} 235 \\ 3\overline{)7077} \\ -6 \\ \hline 10 \\ -9 \\ \hline 17 \\ -15 \\ \hline 27 \end{array} \longrightarrow \begin{array}{r} 2359 \\ 3\overline{)7077} \\ -6 \\ \hline 10 \\ -9 \\ \hline 17 \\ -15 \\ \hline 27 \\ -27 \\ \hline 0 \end{array}$$

4.   256

$$\begin{array}{r} 2 \\ 5\overline{)1280} \\ -10 \\ \hline 28 \end{array} \longrightarrow \begin{array}{r} 25 \\ 5\overline{)1280} \\ -10 \\ \hline 28 \\ -25 \\ \hline 30 \end{array} \longrightarrow$$

$$\longrightarrow \begin{array}{r} 256 \\ 5\overline{)1280} \\ -10 \\ \hline 28 \\ -25 \\ \hline 30 \\ -30 \\ \hline 0 \end{array}$$

5. 5,215

```
      5              52
 9 | 46935  →   9 | 46935   →
  - 45            - 45
  ----            ----
    19              19
                  - 18
                  ----
                    13
```

```
        521             5215
 →  9 | 46935   →   9 | 46935
     - 45             - 45
     ----             ----
       19               19
     - 18             - 18
     ----             ----
       13               13
      - 9              - 9
      ----             ----
        45               45
                       - 45
                       ----
                         0
```

6. 257 R 1

```
      2              25
 2 | 515   →    2 | 515   →
  - 4             - 4
  ----            ----
    11              11
                  - 10
                  ----
                    15
```

```
        257 R1
 →  2 | 515
     - 4
     ----
       11
     - 10
     ----
       15
     - 14
     ----
        1
```

7. 465 R 7

```
      4              46
 17 | 7912  →   17 | 7912   →
   - 68            - 68
   ----            ----
    111             111
                  - 102
                  -----
                     92
```

```
        465 R7
 →  17 | 7912
     - 68
     ----
      111
    - 102
    -----
       92
     - 85
     ----
        7
```

8. 311 R 24

```
      3              31
 25 | 7799  →   25 | 7799   →
   - 75            - 75
   ----            ----
     29              29
                   - 25
                   ----
                     49
```

```
        311 R24
 →  25 | 7799
     - 75
     ----
       29
     - 25
     ----
       49
     - 25
     ----
       24
```

9. 1,253 R 41

$$
\begin{array}{r} 1 \\ 43 \overline{)53920} \\ -43 \\ \hline 109 \end{array}
\longrightarrow
\begin{array}{r} 12 \\ 43 \overline{)53920} \\ -43 \\ \hline 109 \\ -86 \\ \hline 232 \end{array}
\longrightarrow
$$

$$
\longrightarrow
\begin{array}{r} 125 \\ 43 \overline{)53920} \\ -43 \\ \hline 109 \\ -86 \\ \hline 232 \\ -215 \\ \hline 170 \end{array}
\longrightarrow
\begin{array}{r} 1253\ R41 \\ 43 \overline{)53920} \\ -43 \\ \hline 109 \\ -86 \\ \hline 232 \\ -215 \\ \hline 170 \\ -129 \\ \hline 41 \end{array}
$$

10. 956 R 76

$$
\begin{array}{r} 9 \\ 128 \overline{)122444} \\ -1152 \\ \hline 724 \end{array}
\longrightarrow
\begin{array}{r} 95 \\ 128 \overline{)122444} \\ -1152 \\ \hline 724 \\ -640 \\ \hline 844 \end{array}
\longrightarrow
$$

$$
\longrightarrow
\begin{array}{r} 956\ R76 \\ 128 \overline{)122444} \\ -1152 \\ \hline 724 \\ -640 \\ \hline 844 \\ -768 \\ \hline 76 \end{array}
$$

☞ Practice: Whole Numbers

DIRECTIONS: Answer each question as indicated.

1. The place value of 3 in the number 123,498 is _____.

2. 7,030,209 written in words is _____.

3. The number 640,018 when rounded to the tens position is _____.

4. The number five hundred ten thousand, seventy three written with digits appears as _____.

5. The number 4,895,121 when rounded to the ten thousands position is _____.

6. 21,868 _____ 21,768. Fill in the blank with <, =, or >.

7. $4,996 + 180 + 3,072 = $ _____.

8. $2,017 - 1,349 = $ _____.

9. $703 \times 24 = $ _____.

10. $8,291 \times 103 = $ _____.

11. $1,985 \div 3 = $ _____.

12. $56,061 / 9 = $ _____.

13. $263,832 / 15 = $ _____.

14. $37,772 / 133 = $ _____.

15. The digit corresponding to the hundred thousands place in the number 12,563,047 is _____.

16. 78,003 written in words is _____.

17. The number 53,647 when rounded to the hundreds position is _____.

18. The number two hundred eight million, fifteen thousand, nine hundred written with digits appears as _____.

19. 563,281 _____ 563,194. Fill in the blank with <, =, or >.

20. The number 809,956,111 when rounded to the hundred thousands position is _____.

21. 5,581 + 10,617 + 92,003 = _____.

22. 210,514 – 83,623 = _____.

23. 925 × 259 = _____.

24. 5,194 / 19 = _____.

25. 11,988 ÷ 324 = _____.

Answers

1. Thousands. It is the fourth digit to the left of the decimal point.

2. Seven million, thirty thousand, two hundred nine.

3. 640,020. Since the units digit 8 is greater than or equal to 5, the tens digit is raised by 1. The units digit becomes 0.

4. 510,073

5. 4,900,000. Since the thousands digit is 5, raise the ten thousands digit by 1. The 9 becomes a 0 and the 8 becomes a 9. All digits to the right of the 9 become 0's.

6. > because the first difference from left to right shows 8 compared to 7 in the hundreds place.

7. 8,248

$$\begin{array}{r} \underline{1\ 2} \\ 4996 \\ 180 \\ +\ 3072 \\ \hline 8248 \end{array}$$

8. 668

$$\begin{array}{r} \underline{9\ 10\ 1} \\ 2017 \\ -\ 1349 \\ \hline 668 \end{array}$$

9. 16,872

$$\begin{array}{r} 703 \\ \times\ 24 \\ \hline 2812 \\ +\ 1406 \\ \hline 16872 \end{array}$$

10. 853,973

$$\begin{array}{r} 8291 \\ \times\ 103 \\ \hline 24873 \end{array} \longrightarrow \begin{array}{r} 8291 \\ \times\ 103 \\ \hline 24873 \\ +\ 82910 \\ \hline 853973 \end{array}$$

11. 661 R 2

```
      6              66
 3 ⟌1985   →   3 ⟌1985   →
  − 18          − 18
    18            18
                − 18
                   0
```

```
              661 R2
   →     3 ⟌1985
          − 18
            18
          − 18
            05
           − 3
             2
```

12. 6,229

```
      6              62
 9 ⟌56061  →   9 ⟌56061  →
  − 54          − 54
    20            20
                − 18
                  26
```

```
            622              6229
  →   9 ⟌56061   →   9 ⟌56061   →
       − 54            − 54
         20              20
       − 18            − 18
         26              26
       − 18            − 18
          8              81
                       − 81
                          0
```

13. 17,588 R 12

```
       1               17
15 ⟌263832   →   15 ⟌263832   →
  − 15            − 15
    11             113
                 − 105
                    88
```

```
           175                1758
  →   15 ⟌263832   →   15 ⟌263832   →
       − 15               − 15
        113                113
      − 105              − 105
         88                 88
       − 75               − 75
         13                133
                        − 120
                           13
```

```
            17588 R12
  →   15 ⟌263832
       − 15
        113
      − 105
         88
       − 75
        133
      − 120
        132
      − 120
         12
```

14. 284

$$
\begin{array}{r}
2 \\
133\overline{)37772} \\
-266 \\
\hline
111
\end{array}
\longrightarrow
\begin{array}{r}
28 \\
133\overline{)37772} \\
-266 \\
\hline
1117 \\
-1064 \\
\hline
53
\end{array}
\longrightarrow
$$

$$
\longrightarrow
\begin{array}{r}
284 \\
133\overline{)37772} \\
-266 \\
\hline
1117 \\
-1064 \\
\hline
532 \\
-532 \\
\hline
0
\end{array}
$$

15. 5. This is the sixth digit counting from right to left.

16. Seventy eight thousand, three.

17. 53,600. Since the tens digit 4 is less than 5, leave the 6 as is. Change the 4 and 7 to 0's.

18. 208,015,900

19. > because the first difference from left to right shows 2 compared to 1 in the hundreds place.

20. 810,000,000. The digit 5 is 5 or higher, so the 9 in the hundred thousands position goes to a 0. This forces the 9 in the millions position to become 0 and thus force the 0 in the ten millions position to become 1. All digits to the right of 5 and including 5 become 0's.

21. 108,201

$$
\begin{array}{r}
\mathit{111} \\
5581 \\
10617 \\
+\ 92003 \\
\hline
108201
\end{array}
$$

22. 126,891

$$
\begin{array}{r}
\mathit{0\,9\,14\,1} \\
210514 \\
-\ 83623 \\
\hline
126891
\end{array}
$$

23. 239,575

$$
\begin{array}{r}
925 \\
\times 259 \\
\hline
8325
\end{array}
\longrightarrow
\begin{array}{r}
925 \\
\times 259 \\
\hline
8325 \\
+\ 4625
\end{array}
\longrightarrow
\begin{array}{r}
925 \\
\times 259 \\
\hline
8325 \\
4625 \\
+\ 1850 \\
\hline
239575
\end{array}
$$

24. 273 R 7

$$
\begin{array}{r}
2 \\
19\overline{)5194} \\
-38 \\
\hline
13
\end{array}
\longrightarrow
\begin{array}{r}
27 \\
19\overline{)5194} \\
-38 \\
\hline
139 \\
-133 \\
\hline
6
\end{array}
\longrightarrow
$$

$$
\longrightarrow
\begin{array}{r}
273\ \text{R7} \\
19\overline{)5194} \\
-38 \\
\hline
139 \\
-133 \\
\hline
64 \\
-57 \\
\hline
7
\end{array}
$$

25. 37

$$324 \overline{|11988} \quad \longrightarrow \quad 324 \overline{|11988}$$

$$\begin{array}{r} 3 \\ 324 \overline{\smash{)}11988} \\ -972 \\ \hline 226 \end{array} \quad \longrightarrow \quad \begin{array}{r} 3 \\ 324 \overline{\smash{)}11988} \\ -972 \\ \hline 2268 \\ -2268 \\ \hline 0 \end{array}$$

REVIEW

In this section, you learned the value of whole numbers and their use in everyday life. It is important to know how to identify the place values of digits in whole numbers. By understanding the different place values (ones, tens, hundreds, thousands, ten thousands, etc.), you will be able to convert words to numbers and vice versa. For example, three thousand nine hundred twenty-six would be written out as 3,926. This skill has many practical applications in everyday life, such as writing out a check, or taking an inventory, for instance.

Recognizing the size (or value) of a number when two or more numbers have an equal amount of digits is also important. Keep in mind that if one number has more digits, it is the larger number. By remembering this simple rule, you can save yourself time in figuring out price information at the store, or when solving mathematical problems in class.

When rounding off to a place for any whole number, be sure to check the digit to the right of the desired rounding number; if it is five or higher, raise the digit in the desired location by one. If not, leave that digit alone. For instance, if we wanted to round 158 to the tens place, the correct answer would be 160 because 8 (in the ones place) is higher than 5. This skill is directly related to identifying different place values in numbers and is often used for making quick approximations.

Adding, subtracting, multiplying, and dividing whole numbers are functions that people use nearly every day throughout their lives. When doing simple addition and subtraction of whole numbers, be sure to correctly align the different place values (ones, tens, hundreds, etc.). This will ensure that the proper answer is reached. When multiplying whole numbers, be sure to move from right to left, starting with the right-most digit of the first number. To divide whole numbers, however, move from left to right. Be sure to begin with the divisor and determine the number of times it can be divided into the appropriate number of digits in the dividend. Be aware of any remainders, and keep in mind that remainders must be less than the divisor.

These basic mathematical skills of writing numbers, rounding numbers, addition, subtraction, multiplication, and division are used in everyday life in situations ranging from renting movies at the video store, writing checks to pay bills, to dealing with all monetary transactions. Once these simple skills have been mastered, you will be able to move on to more advanced calculations, using what you have just learned.

Mathematics

Fractions

MATHEMATICS

FRACTIONS

INTRODUCTION TO FRACTIONS

As we saw a few sections ago, whole numbers are just one type of number that can be discussed. We saw that decimals can be thought of as "pieces" of whole numbers. Another way of thinking of these "pieces" is fractions. Again, we are looking at "parts" of a number. An example of a fraction would be 1/2. Instead of dividing out the numbers and getting a decimal, we leave it in a "fraction format." The number on the top part of the fraction (1 in this case) is referred to as the **numerator**. The bottom part of a fraction (2 in this example) is called the **denominator**. Depending on the operation that we are looking at, each of these parts of a fraction has its significant role.

Example 1

In the problems below, determine whether the 5 appears in the numerator or denominator.

(a) $\dfrac{3}{5}$

In this example, the 5 appears in the bottom part of the fraction. From our discussion above, we know that this is the denominator.

(b) $\dfrac{5}{7}$

In this example, the 5 appears in the top part of the fraction. From our discussion we know that this is the numerator.

(c) $\dfrac{5}{2}$

In this example, the 5 appears in the top part of the fraction. From our discussion we know that this is the numerator.

(d) $\dfrac{5}{5}$

In this example, the 5 appears in both the top and bottom of the fraction. From our discussion we know that here we have 5 appearing in the numerator and denominator.

Any whole number can be made into a fraction by putting it over 1. We can place the whole number in the numerator of a fraction and place a 1 in the denominator. At this point, we might have a hard time seeing why this would be of any importance. But as we proceed through this section, we will come across examples where this will be needed.

Example 2

Write the following whole numbers as fractions.

(a) 3

In order to write this number as a fraction, we can write this value (3) in the numerator and place a one in the denominator. This gives us 3/1.

(b) 7

In order to write this number as a fraction, we can place this value (7) in the numerator and put a one in the denominator. In this example we get 7/1.

(c) 24

In order to write this number as a fraction, we can place this value (24) in the numerator and place a one in the denominator. This gives us 24/1.

(d) 436

In order to write this number as a fraction, we can place this value (436) in the numerator and put a one in the denominator. We get 436/1 for an answer.

If we are considering a fraction where the numerator (top part of the fraction) is smaller than the denominator (bottom part of a fraction), then we have a number that is less than one. Some examples are 1/2, 5/9, and 7/11. In each of the above cases, the numerator is less than the denominator. Some of these fractions can be **reduced**. Instead of working with 3/9, we could consider 1/3. Even though these fractions look different, they have the same value. If we would divide each of them out, we would get exactly the same value (0.3333). How did we do this? By looking at the top and the bottom of 3/9, we see that both the numerator and denominator have a similarity. Both of them can be divided by 3. If we divide 3 by 3 and 9 by 3, we would get 1 and 3. These are the same values that we stated above.

$$\frac{3}{9} \begin{array}{c} \rightarrow 3/3 = 1 \rightarrow \\ \rightarrow 9/3 = 3 \rightarrow \end{array} \frac{1}{3}$$

If we have any fractions where the top and bottom have this type of similarity, we know that we can reduce them. By reducing a fraction, we are rewriting it with smaller values in the numerator and denominator.

Example 3

Reduce the following fractions.

(a) $\frac{4}{8}$

In this example, we have a 4 in the numerator and an 8 in the denominator. We can see that they are both divisible by 4. If we divide the numerator and denominator by 4, we get 1/2. Below are the detailed steps.

$$\frac{4}{8} \begin{array}{c} \rightarrow 4/4 = 1 \rightarrow \\ \rightarrow 8/4 = 2 \rightarrow \end{array} \frac{1}{2}$$

(b) $\frac{5}{20}$

In this example, we have a 5 in the numerator and a 20 in the denominator. We can see that they are both divisible by 5. If we divide the numerator and denominator by 5, we get 1/4. Below are the steps needed for this reduction.

$$\frac{5}{20} \begin{array}{c} \rightarrow 5/5 = 1 \rightarrow \\ \rightarrow 20/5 = 4 \rightarrow \end{array} \frac{1}{4}$$

(c) $\frac{6}{9}$

In this example, we have a 6 in the numerator and a 9 in the denominator. We can see that they are both divisible by 3. If we do this division, we get 2/3. The necessary steps follow.

$$\frac{6}{9} \longrightarrow \begin{matrix} 6/3 = 2 \\ 9/3 = 3 \end{matrix} \longrightarrow \frac{2}{3}$$

$$\frac{2}{2} \longrightarrow \begin{matrix} 2/2 = 1 \\ 2/2 = 1 \end{matrix} \longrightarrow \frac{1}{1} = 1$$

(d) $\dfrac{20}{25}$ $\dfrac{4}{5}$

In this example, we have a 20 in the numerator and a 25 in the denominator. We can see that they are both divisible by 5. This gives 4/5. Below are the details needed to get this answer.

$$\frac{20}{25} \longrightarrow \begin{matrix} 20/5 = 4 \\ 25/5 = 5 \end{matrix} \longrightarrow \frac{4}{5}$$

When we have a fraction where the numerator *and* denominator are the same value, we can rewrite this fraction as 1. In order to see this in more detail, we need to use our understanding from Examples 2 and 3 above. We first reduce the fraction into the form 1/1. We can then rewrite this value as the single whole number 1. By considering the problems below, we will get a better understanding of how this process works.

Example 4

Simplify the following fractions.

(a) $\dfrac{2}{2}$

In this example, we have a 2 in the numerator and denominator. These values can be divided by 2. This gives us 1/1. Since any whole number can be written as a fraction by placing it over one, we can rewrite this fraction as 1. In the next column are the details needed to get this answer.

(b) $\dfrac{5}{5}$

In this example, we have a 5 in the numerator and denominator. These values can be divided by 5. This gives us 1/1. Since any whole number can be written as a fraction by placing it over one, we can rewrite this fraction as 1. Below are the details needed to get this answer.

$$\frac{5}{5} \longrightarrow \begin{matrix} 5/5 = 1 \\ 5/5 = 1 \end{matrix} \longrightarrow \frac{1}{1} = 1$$

(c) $\dfrac{12}{12}$

In this example, we have a 12 in the numerator and denominator. These values can be divided by 12. This gives us 1/1. Since any whole number can be written as a fraction by placing it over one, we can rewrite this fraction as 1. Below are the details needed to get this answer.

$$\frac{12}{12} \longrightarrow \begin{matrix} 12/12 = 1 \\ 12/12 = 1 \end{matrix} \longrightarrow \frac{1}{1} = 1$$

(d) $\dfrac{86}{86}$

In this example, we have an 86 in the numerator and denominator. These values can be divided by 86. This gives us 1/1. Since any whole number can be written as a fraction by placing it over one, we can rewrite this fraction as 1. On the next page are the details needed to get this answer.

$$\frac{86}{86} \xrightarrow{\longrightarrow} \begin{array}{c} 86/86 = 1 \xrightarrow{\longrightarrow} 1 \\ 86/86 = 1 \xrightarrow{\longrightarrow} 1 \end{array} = 1$$

If the numerator is greater than the denominator, we have a fraction greater than one. These types of numbers are also referred to as **improper fractions** and can be rewritten as **mixed numbers**. A mixed number is a whole number combined with a fraction. Instead of writing the number 11/6, we can also think of it as 1⅚. At first glance this seems strange. But, as we considered above, when the numerator and denominator match, we get 1. If we subtract 6 from 11, we get 5. This accounts for the 5 appearing as the numerator of the fractional part of the mixed number. Notice that the denominator of the fractional part has the same value as the denominator of the original fraction. In the example below, we will discuss converting improper fractions into mixed numbers.

There is another way of approaching improper fractions that some students may find easier.

Take 11/6 from the above example. What we want to do here is take the denominator of the improper fraction and divide it into the numerator. In other words, we want to see how many 6's are in 11. We see that 6 goes into 11 once, with a remainder of 5 (or there is only one 6 in 11 with 5 left over). We then place the 5 over the original denominator, 6, and place the 1 (for the amount of times 6 goes into 11) in front of 5/6, to get the mixed number 1⅚.

Example 5

Convert these fractions into mixed numbers.

(a) $\dfrac{7}{4}$

In this example, we can see that the numerator is larger than the denominator. This tells us that this fraction has a value larger than 1. By subtracting the denominator from the numerator (7 − 4 = 3), we will get the numerator of the fractional part of our mixed number. Therefore, another way of saying 7/4 is 1¾.

If we divide, we get 7 is divisible by 4 once with 3 left over or 1¾.

(b) $\dfrac{9}{7}$

In this example, we can see that the numerator is larger than the denominator. This tells us that this fraction has a value larger than 1. By subtracting the denominator from the numerator (9 − 7 = 2), we will get the numerator of the fractional part of our mixed number. Therefore, another way of saying 9/7 is 1²/₇.

Dividing 9 by 7 will give us 1, with 2 left over, or 1²/₇.

(c) $\dfrac{10}{6}$

In this example, we can see that the numerator is larger than the denominator. This tells us that this fraction has a value larger than 1. By subtracting the denominator from the numerator (10 − 6 = 4), we will get the numerator of the fractional part of our mixed number. Therefore, another way of saying 10/6 is 1⁴/₆. Can this answer be simplified more? Consider the fractional part (4/6). This can be reduced using the properties that we discussed in earlier sections. That is,

$$\frac{4}{6} \begin{array}{c} \longrightarrow 4/2 = 2 \longrightarrow \\ \longrightarrow 6/2 = 3 \longrightarrow \end{array} \frac{2}{3}$$

Therefore, our final answer in reduced form would be 1²/₃. Another way of solving this problem would be to reduce the original improper fraction and then convert that into a mixed number. Either way, we would get the same answer.

(d) $\frac{11}{5}$

In this example, we can see that the numerator is larger than the denominator. This tells us that this fraction has a value larger than 1. By subtracting the denominator from the numerator (11 − 5 = 6), we will get the numerator of the fractional part of our mixed number. Therefore, another way of saying 11/5 is 1⁶/₅. Can this answer be simplified more? Consider the fractional part (6/5). This is still an improper fraction. This tells us that we are still working with a fraction larger than 1. By subtracting the denominator from the numerator (6 − 5 = 1), we will get the numerator of the fractional part of our mixed number. Therefore, another way of saying 1⁶/₅ is 2¹/₅.

Since we are able to rewrite improper fractions as mixed numbers, it would make sense that we could also go the other way. Changing mixed numbers into improper fractions is very similar to the above discussion, but in reverse. The next example will demonstrate how this can be done.

Example 6

Write the following mixed number as an improper fraction.

(a) $1\frac{1}{2}$

Before we subtracted the denominator from the numerator to get the fractional part of our answer. In this case, we want to do the opposite. Here we want to think of the whole number 1 as 2/2. We know that these values are equal. Therefore, by adding the numerator and denominator together (1 + 2 = 3), we will get the numerator of our improper fraction. Another way of saying 1¹/₂ would be 3/2. Notice that the denominator of the fractional part becomes the denominator of the improper fraction.

Another way to get this answer is to multiply the denominator by the whole number and add that sum to the numerator to get our new numerator. Leaving out the whole number, and keeping the original denominator, we get the improper fraction. Take 1¹/₂ from above. We multiply 2 and 1 and get 2. We then add this number to our current numerator to get 3. Placing the 3 in the numerator and the 2 in the denominator, we have 3/2.

(b) $1\frac{3}{8}$

Here we want to think of the 1 as 8/8. We know that these values are equal. Therefore, by adding the numerator and denominator together (3 + 8 = 11), we will get the numerator of our improper fraction. Another way of saying $1\frac{3}{8}$ would be 11/8. Again, notice that the denominator of the fractional part becomes the denominator of the improper fraction.

(c) $1\frac{7}{11}$

Here we want to think of the 1 as 11/11. We know that these values are equal. Therefore, by adding the numerator and denominator together (7 + 11 = 18), we will get the numerator of our improper fraction. Another way of saying $1\frac{7}{11}$ would be 18/11. Again, the denominator of the fractional part becomes the denominator of the improper fraction.

(d) $2\frac{4}{5}$

In this case, we are dealing with a number that has a whole part larger than 1. In order to handle this problem, we can apply the above procedure twice. Each of these procedures will convert one whole number into the corresponding fractional part. First, we want to think of the 2 as 5/5. We know that these values are equal. Therefore, by adding the numerator and denominator together (4 + 5 = 9), we will get the numerator of our improper fraction. Another way of saying $2\frac{4}{5}$ would be $1\frac{9}{5}$. We can replace the remaining one with its fraction part. We again think of 1 as 5/5. By adding the numerator and denominator together (9 + 5 = 14), we will get the numerator of our improper fraction. Another way of saying $1\frac{9}{5}$ would be 14/5. Therefore, we can write $2\frac{4}{5}$ as 14/5. Notice that the denominator of the

fractional part becomes the denominator of the improper fraction.

In previous sections, we discussed the ideas of a number being less than another, two numbers being equal, and a number being greater than another. In the next section we will also discuss this notion using fractions.

Questions

In the fractions below, determine whether the even number appears in the numerator or denominator.

1. $\frac{7}{8}$

2. $\frac{10}{13}$

3. $\frac{4}{6}$

4. $\frac{5}{9}$

5. $\frac{12}{5}$

Write the numbers below as fractions.

6. 4

7. 7

8. 19

9. 43

10. 152

Reduce the following fractions as much as possible.

11. $\frac{10}{15}$

12. $\frac{3}{9}$

13. $\frac{27}{49}$

14. $\frac{12}{18}$

15. $\frac{30}{60}$

Simplify the following fractions.

16. $\frac{9}{9}$

17. $\frac{55}{55}$

18. $\frac{14}{14}$

19. $\frac{4}{4}$

20. $\frac{634}{634}$

Convert these fractions into mixed numbers.

21. $\frac{8}{5}$

22. $\frac{15}{9}$

23. $\frac{26}{19}$

24. $\frac{18}{7}$

25. $\frac{15}{4}$

Write the following mixed numbers as improper fractions.

26. $1\frac{2}{5}$

27. $1\frac{8}{9}$

28. $1\frac{12}{17}$

29. $2\frac{3}{14}$

30. $3\frac{1}{2}$

Answers

1. Denominator

2. Numerator

3. Both—numerator and denominator

4. Neither—numerator nor denominator

5. Numerator

6. $\dfrac{4}{1}$

7. $\dfrac{7}{1}$

8. $\dfrac{19}{1}$

9. $\dfrac{43}{1}$

10. $\dfrac{152}{1}$

11. $\dfrac{2}{3}$

$$\dfrac{10 \,(\div 5)}{15 \,(\div 5)} = \dfrac{2}{3}$$

12. $\dfrac{1}{3}$

$$\dfrac{3 \,(\div 3)}{9 \,(\div 3)} = \dfrac{1}{3}$$

13. $\dfrac{27}{49}$ Cannot be reduced.

14. $\dfrac{2}{3}$

$$\dfrac{12 \,(\div 6)}{18 \,(\div 6)} = \dfrac{2}{3}$$

15. $\dfrac{1}{2}$

$$\dfrac{30 \,(\div 30)}{60 \,(\div 30)} = \dfrac{1}{2}$$

16. 1

17. 1

18. 1

19. 1

20. 1

21. $1\dfrac{3}{5}$

$$\dfrac{8}{5} = 1\dfrac{3}{5}$$

22. $1\dfrac{2}{3}$

$$\dfrac{15}{9} = 1\dfrac{2}{3}$$

23. $1\dfrac{7}{19}$

$$\dfrac{26}{19} = 1\dfrac{7}{19}$$

24. $2\dfrac{4}{7}$

$$\dfrac{18}{7} = 2\dfrac{4}{7}$$

25. $3\dfrac{3}{4}$

$$\dfrac{15}{4} = 3\dfrac{3}{4}$$

26. $\dfrac{7}{5}$

$$\dfrac{7}{5} = 1\dfrac{2}{5}$$

27. $\dfrac{17}{9}$

$$\dfrac{17}{9} = 1\dfrac{8}{9}$$

28. $\dfrac{29}{17}$

$$\dfrac{29}{17} = 1\dfrac{12}{17}$$

29. $\dfrac{31}{14}$

$$\dfrac{31}{14} = 2\dfrac{3}{14}$$

30. $\dfrac{7}{2}$

$$\dfrac{7}{2} = 3\dfrac{1}{2}$$

ADDITION OF FRACTIONS

Now that we understand what fractions and mixed numbers are, we want to discuss adding, subtracting, multiplying, and dividing. This section looks at addition. In the following sections, we will address the other operations. We will discover, as we have in previous sections, that working with fractions has similarities to working with whole numbers. These topics are not exactly the same, but we will see how a good understanding of whole numbers will help us here.

The first case of adding fractions that we will consider will be when the denominators are equal. In this case, we just need to add the numerators together and bring the common denominator along for the ride. As we saw in our discussion on mixed numbers, the denominator will play a very important role in addition.

Example 1

Add the following fractions.

(a) $\dfrac{1}{3} + \dfrac{1}{3}$

In this example, we notice that the denominators of these fractions have the same value (3). This being the case, we just need to add the numerators together (1 + 1 = 2) and bring the denominator (3) along for the ride. Therefore, our final answer is 2/3. Below is the detailed step needed to complete this problem.

$$\dfrac{1}{3} + \dfrac{1}{3} = \dfrac{1+1}{3} = \dfrac{2}{3}$$

(b) $\dfrac{1}{5} + \dfrac{3}{5}$

In this example, we notice that the denominators of these fractions have the same value (5). This being the case, we just need to add the numerators together (1 + 3 = 4) and bring the denominator (5) along for the ride. Therefore, our final answer is 4/5. Below is the detailed step needed to complete this problem.

$$\dfrac{1}{5} + \dfrac{3}{5} = \dfrac{1+3}{5} = \dfrac{4}{5}$$

(c) $\dfrac{5}{9}+\dfrac{2}{9}$ $\dfrac{7}{9}$

In this example, we notice that the denominators of these fractions have the same value (9). This being the case, we just need to add the numerators together (5 + 2 = 7) and bring the denominator (9) along for the ride. Therefore, our final answer is 7/9. Below is the detailed step needed to complete this problem.

$$\frac{5}{9}+\frac{2}{9}=\frac{5+2}{9}=\frac{7}{9}$$

(d) $\dfrac{5}{12}+\dfrac{1}{12}$ $\dfrac{6}{12}$ $\dfrac{1}{6}$

In this example, we notice that the denominators of these fractions have the same value (12). This being the case, we just need to add the numerators together (5 + 1 = 6) and bring the denominator (12) along. Therefore, our answer is 6/12. Below is the detailed step needed to complete this problem.

$$\frac{5}{12}+\frac{1}{12}=\frac{5+1}{12}=\frac{6}{12}$$

But is this our final answer? Recall our discussions in the previous section. We saw that some fractions can be reduced. In our answer above, both the numerator and denominator can be reduced by 6. When this is accomplished, our final answer would be 1/2. Below is the detailed step.

$$\frac{6}{12}\longrightarrow\begin{array}{l}6/6=1\\12/6=2\end{array}\longrightarrow\frac{1}{2}$$

We now need to look at problems where the denominators are not the same. In order to add fractions, we need to transform the de-nominator in each fraction into a common value. One way of accomplishing this is by taking the denominators and multiplying them together. We can easily transform each denominator into this value, but in most cases this gives an extremely large value. An example of this would be 6/12 + 7/18. If we multiply the denominators together (12 × 18), we would get 216. This value would work as a common denominator, but is there a smallest value that would also work?

The smallest value possible that could be used is referred to as the **Lowest Common Denominator** or **LCD**. Instead of using large numbers, the LCD allows us to use the smallest value possible. Sometimes this is still a large value, but in most cases, the value we are working with is much more tolerable. The big question now becomes how do we find the LCD. The traditional way to find the LCD is to form factor trees of each denominator and then compare the factors. If we stick with the above problem (6/12 + 7/18), we would break down 12 and 18 into a list of prime numbers multiplied together.

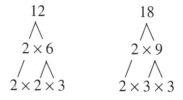

In each level of these factor trees, we still have 12 or 18, depending on which tree we are considering. Also, in each level we are breaking down parts of the original number. We began by breaking 12 into 2 × 6. Since 2 is already prime, we can leave it alone. The 6, on the other hand, is not prime and can be broken down into 2 × 3. By combining everything together, we get 2 × 2 × 3. We followed similar thinking to get the above factor tree for 18.

We now need to compare the last line of each factor tree. We begin by letting the LCD equal the bottom line of the first factor tree. In this example, we would get LCD = $2 \times 2 \times 3$. We now compare each factor in the bottom line of the other factor tree to this LCD and see what is already there and what needs to be included. We first compare the 2 to the LCD that we already have (LCD = $\underline{2} \times 2 \times 3$). Since we already have a two in our LCD line (the underlined 2 in our LCD statement), we do not need to include it again. Now, we compare the next factor (3) to our LCD line. Since we already have a 3 in our LCD line (LCD = $2 \times 2 \times \underline{3}$), we don't need to include it again. The last factor we need to compare to our LCD line is another 3. This time, when we compare the 3 to our LCD line, we see that we have to include another 3 in our LCD line. This is because we have already accounted for the underlined 3. Therefore, our LCD would be $2 \times 2 \times 3 \times 3$ or 36. Even though this value is a little bit large, it is nowhere near the 216 that we discussed. Knowing this, we would then transform each fraction into one that has a 36 in the denominator. Once this is done, we can add the fractions.

Even though this method works, we hope that there is an easier way. Surprisingly, there is yet another method that can be used. The main point of this method is reducing a special fraction called the **test fraction**. This test fraction is made up of the denominators in the original problem. Since denominators play such an active role in adding fractions, we want to get them involved as soon as possible. If we consider the problem from above, 6/12 + 7/18, our test fraction would be 12/18. We just put one denominator on the top and the other on the bottom of our test fraction. Can this fraction be reduced? Our answer here is "Yes!" Both 12 and 18 can be divided by 6 evenly. This gives 2/3.

$$\frac{12}{18} \longrightarrow \begin{array}{c} 12/6 = 2 \\ 18/6 = 3 \end{array} \longrightarrow \frac{2}{3}$$

We can now take this reduced test frac-

tion to get the Lowest Common Denominator (LCD) to appear in both fractions. Since the 2 "corresponds" with the 12 (they are both in the numerator of equal fractions), we want to place this value in both the numerator AND denominator of the fraction that does NOT have 12 in the denominator, that is 7/18. Therefore,

$$\frac{6}{12} + \frac{7}{18} = \frac{6}{12} + \frac{7 \times 2}{18 \times 2}$$

Now we can continue this process by looking at the denominators of the test fraction. Since 3 "corresponds" with the 18 (they are both in the denominator of equal fractions), we want to place this value in both the numerator AND denominator of the fraction that does NOT have 18 in the denominator, that is, 6/12. Therefore,

$$\frac{6}{12} + \frac{7 \times 2}{18 \times 2} = \frac{6 \times 3}{12 \times 3} + \frac{7 \times 2}{18 \times 2}$$

At first glance, it does not seem to help in any way. But, if we simplify both of these fractions, we will get

$$\frac{6 \times 3}{12 \times 3} + \frac{7 \times 2}{18 \times 2} = \frac{18}{36} + \frac{14}{36}$$

At this point, we notice the denominators are the same. The surprising thing is that this value (36) is the LCD. Even though this method appears lengthy, all we have to do is reduce our test fraction. Here, we don't need to worry about factor trees (or anything like this). Now that the denominators are the same, we can add the numerators. If we complete this problem, we will get

$$\frac{18}{36} + \frac{14}{36} = \frac{18 + 14}{36} = \frac{32}{36}$$

Even though we have simplified these fractions, this is not our final answer. As with our test fraction, this answer can also be reduced.

$$\frac{32}{36} \longrightarrow \frac{32/4}{36/4} \longrightarrow \frac{8}{9}$$

At last, we have our answer. Thus, 6/12 + 7/18 = 8/9. Both methods discussed above will always work. Since most students prefer the second one, this will be the one stressed here.

Example 2

Add the following fractions.

(a) $\dfrac{3}{10} + \dfrac{2}{15}$

Our test fraction in this example is 10/15. This is a must here because the denominators do not match. Our test fraction can be reduced to find out what value must be used for the LCD.

$$\frac{10}{15} \longrightarrow \frac{10/5 = 2}{15/5 = 3} \longrightarrow \frac{2}{3}$$

Since the 2 corresponds with the 10, we multiply the numerator and the denominator of the fraction that does not contain the 10 in the denominator by 2.

$$\frac{3}{10} + \frac{2}{15} = \frac{3}{10} + \frac{2 \times 2}{15 \times 2}$$

Likewise, since 3 corresponds to 15, we multiply the numerator and the denominator of the fraction that does not contain the 15 in the denominator by 3.

$$\frac{3}{10} + \frac{2 \times 2}{15 \times 2} = \frac{3 \times 3}{10 \times 3} + \frac{2 \times 2}{15 \times 2}$$

We simplify both of these fractions and see that the denominators now match. This being the case, we are able to add the numerators together.

$$\frac{9}{30} + \frac{4}{30} = \frac{9+4}{30} = \frac{13}{30}$$

Since this answer cannot be reduced or rewritten as a mixed number, we are done with this problem.

(b) $\dfrac{1}{2} + \dfrac{3}{4}$

In order to add these fractions, the bottom part of each fraction must be the same value. Since this is not the case, we can use our test fraction idea to find out what the LCD (or Lowest Common Denominator) must be. In this example, our test fraction would be 2/4. We can reduce this fraction and get 1/2.

$$\frac{2}{4} \longrightarrow \frac{2/2 = 1}{4/2 = 2} \longrightarrow \frac{1}{2}$$

Since the 1 corresponds to 2 (they are both in the numerator of equal fractions), we want to multiply the numerator and denominator of the other fraction (the one that contains the 4 in the bottom) by 1.

$$\frac{1}{2} + \frac{3}{4} = \frac{1}{2} + \frac{3 \times 1}{4 \times 1}$$

We can now continue this process by comparing the denominators of the test fraction. Since 2 corresponds to 4, we want to multiply the other fraction (top and bottom) by 2.

$$\frac{1}{2} + \frac{3 \times 1}{4 \times 1} = \frac{1 \times 2}{2 \times 2} + \frac{3 \times 1}{4 \times 1}$$

If we simplify both of these fractions, we get 2/4 + 3/4. Finally, we come to a point where the bottoms of the fractions match. This being the case, we can add the tops.

$$\frac{2}{4} + \frac{3}{4} = \frac{2+3}{4} = \frac{5}{4}$$

As we discussed earlier, anytime that the top of a fraction is larger than the bottom, we are working with a number larger than one. Therefore, our answer can be written as a mixed number. If we subtract the denominator from the numerator (5 – 4 = 1), we see that 5/4 is the same as 1¼.

(c) $\quad \dfrac{5}{6} + \dfrac{3}{8}$

In order to add these fractions, the denominator must be the same value. To begin with they are not, but we can create our test fraction and use it to determine what value should be used. Here, our test fraction is 6/8.

$$\frac{6}{8} \begin{array}{c} \longrightarrow 6/2 = 3 \longrightarrow \\ \longrightarrow 8/2 = 4 \longrightarrow \end{array} \frac{3}{4}$$

Since 3 "corresponds" with the 6 (they are both in the numerator of equal fractions), we want to place this value in both the numerator **and** denominator of the fraction that does **not** have 6 in the denominator. That is, the fraction that has the 8.

$$\frac{5}{6} + \frac{3}{8} = \frac{5}{6} + \frac{3 \times 3}{8 \times 3}$$

Since 4 "corresponds" with the 8 (they are both in the denominator of equal fractions), we want to place this value in both the numerator **and** denominator of the fraction that does **not** have 8 in the denominator. That is, the fraction that has the 6.

$$\frac{5}{6} + \frac{3 \times 3}{8 \times 3} = \frac{5 \times 4}{6 \times 4} + \frac{3 \times 3}{8 \times 3}$$

If we now simplify both of these fractions, we will get 20/24 + 9/24. Since the denominators are finally the same, we can add the two fractions. This would give

$$\frac{20}{24} + \frac{9}{24} = \frac{20+9}{24} = \frac{29}{24}$$

Since our numerator is larger than the denominator, we know that this answer is larger than one and can be written as a mixed number. By subtracting the denominator from the numerator (29 – 24 = 5), we will get the numerator of the fractional part of our mixed number. Therefore, another way of saying 29/24 is 1⁵⁄₂₄.

(d) $\quad \dfrac{1}{3} + \dfrac{2}{5}$

As we have seen above, we must begin by getting the denominators of these fractions to match. Our idea of a test fraction can still be used here, but we need to be very careful of our steps. This example is a little tricky! Here our test fraction is 3/5. If we try to reduce this fraction, we see that it is impossible. So, where do we go from here? Let's try the same ideas that we used above and see if they will still work here.

$$\frac{3}{5} \begin{array}{c} \longrightarrow 3/? = ? \longrightarrow \\ \longrightarrow 5/? = ? \longrightarrow \end{array} \frac{3}{5}$$

Since the 3 corresponds to the 3, we can multiply the numerator and denominator of the fraction that contains the 5 in the denominator by 3.

$$\frac{1}{3} + \frac{2}{5} = \frac{1}{3} + \frac{2 \times 3}{5 \times 3}$$

Since the 5 corresponds with the 5, we can multiply the numerator and denominator of the fraction that contains the 3 in the denominator by 5.

$$\frac{1}{3} + \frac{2 \times 3}{5 \times 3} = \frac{1 \times 5}{3 \times 5} + \frac{2 \times 3}{5 \times 3}$$

If we simplify both of these fractions, we see that we get the LCD of 15 to appear in the denominator.

$$\frac{5}{15} + \frac{6}{15} = \frac{5 + 6}{15} = \frac{11}{15}$$

This fraction cannot be reduced or rewritten as a mixed number. So, we are done with this example. Notice that even though our test fraction did not simplify as the others did, we can still use this idea to complete our problem.

Now that we have an understanding of a Lowest Common Denominator, or LCD, we want to consider the order of fractions. As we saw in previous sections, two numbers can compare in one of three ways: <, =, and >. When working with fractions, the easiest way to approach this discussion is to transform both fractions into fractions with the same denominator, and then compare the numerators. Once both fractions have the same denominator, the larger numerator would denote the larger fraction. Since our test fraction idea is used to make two denominators match, it seems proper to use it again here.

Example 3

Compare the following pair of fractions. Write the appropriate symbol between the two numbers.

(a) $\dfrac{1}{4}$ _____ $\dfrac{1}{6}$

If we construct our test fraction

(4/6), we can see that it can be reduced to 2/3. We can now multiply the numerator and denominator of the fraction that contains the 6 in the denominator by 2. Likewise, we can multiply the numerator and denominator of the fraction that contains the 4 in the denominator by 3.

$$\frac{1}{4} = \frac{1 \times 3}{4 \times 3} = \frac{3}{12} \quad \text{and} \quad \frac{1}{6} = \frac{1 \times 2}{6 \times 2} = \frac{2}{12}$$

Therefore, this problem can be recast as 3/12 _____ 2/12. Since 3 is larger than 2, we know 3/12 (or 1/4) is a larger fraction. Thus, the proper symbol to be placed between these fractions is ">."

$$\frac{3}{12} > \frac{2}{12} \quad \text{or} \quad \frac{1}{4} > \frac{1}{6}$$

(b) $\dfrac{1}{7}$ _____ $\dfrac{3}{14}$

If we construct our test fraction (7/14), we can see that it can be reduced to 1/2. We can now multiply the numerator and denominator of the fraction that contains the 14 in the denominator by 1. Likewise, we can multiply the numerator and denominator of the fraction that contains the 7 in the denominator by 2.

$$\frac{1}{7} = \frac{1 \times 2}{7 \times 2} = \frac{2}{14} \quad \text{and}$$

$$\frac{3}{14} = \frac{3 \times 1}{14 \times 1} = \frac{3}{14}$$

Therefore, this problem can be recast as 2/14 _____ 3/14. Since 3 is larger than 2, we know 3/14 is a larger fraction. Thus, the proper symbol to be placed between these fractions is "<."

$$\frac{2}{14} < \frac{3}{14} \quad \text{or} \quad \frac{1}{7} < \frac{3}{14}$$

(c) $\dfrac{7}{25} \rule{2cm}{0.4pt} \dfrac{3}{20}$

If we construct our test fraction (25/20) we can see that it can be reduced to 5/4. We can now multiply the numerator and denominator of the fraction that contains the 20 in the denominator by 5. Likewise, we can multiply the numerator and denominator of the fraction that contains the 25 in the denominator by 4.

$$\frac{7}{25} = \frac{7 \times 4}{25 \times 4} = \frac{28}{100} \quad \text{and}$$

$$\frac{3}{20} = \frac{3 \times 5}{20 \times 5} = \frac{15}{100}$$

Therefore, this problem can be recast as 28/100 _____ 15/100. Since 28 is larger than 15, we know 28/100 (or 7/25) is a larger fraction. Thus, the proper symbol to be placed between these fractions is ">."

$$\frac{28}{100} > \frac{15}{100} \quad \text{or} \quad \frac{7}{25} > \frac{3}{20}$$

(d) $\dfrac{2}{4} \rule{2cm}{0.4pt} \dfrac{3}{6}$

If we construct our test fraction (4/6) we can see that it can be reduced to 2/3. We can now multiply the numerator and denominator of the fraction that contains the 6 in the denominator by 2. Likewise, we can multiply the numerator and denominator of the fraction that contains the 4 in the denominator by 3.

$$\frac{2}{4} = \frac{2 \times 3}{4 \times 3} = \frac{6}{12} \quad \text{and}$$

$$\frac{3}{6} = \frac{3 \times 2}{6 \times 2} = \frac{6}{12}$$

Therefore, this problem can be recast as 6/12 _____ 6/12. Since 6 equals 6, we know these numbers are equal to each other. Thus, the proper symbol to be placed between these fractions is "=."

$$\frac{6}{12} = \frac{6}{12} \quad \text{or} \quad \frac{2}{4} = \frac{3}{6}$$

Another approach to this problem would be to notice that both of the original fractions can be reduced. If this was done, we would see that both fractions equal 1/2. Still another way of approaching these problems would be to divide them out and compare the decimal values.

Now that we are "experts" at adding fractions, we want to consider adding mixed numbers. As we saw earlier in this section, we can rewrite mixed numbers as improper fractions. With this in mind, we want to convert the mixed numbers into fractions, add the fractions as we did above, and then convert our answer back into a mixed number if needed. The examples below show this idea in more detail.

Example 4

Add the following numbers.

(a) $1\dfrac{1}{2} + 1\dfrac{2}{3}$

We can begin by converting both of these mixed numbers into improper fractions. This would change the problem into 3/2 + 5/3. Our test fraction in this example is 2/3. Since we cannot reduce this fraction, we have 2/3 = 2/3. We now know that we can multiply the top and bottom of the fraction containing a 3 in the denominator by 2. Likewise, we can multiply the top and bottom of the fraction containing a 2 in the denominator by 3.

$$\frac{3}{2}+\frac{5}{3}=\frac{3\times 3}{2\times 3}+\frac{5\times 2}{3\times 2}=\frac{9}{6}+\frac{10}{6}$$

Now that we have matching denominators, we can add the numerators. This gives us the following:

$$\frac{9}{6}+\frac{10}{6}=\frac{9+10}{6}=\frac{19}{6}$$

Again, in this example, our answer is an improper fraction. We know this since the top of our fraction is larger than the bottom. If we subtract the denominator from the numerator (19 − 6 = 13), we see that 19/6 equals $1^{13}/_6$. We can continue this process a few more times and get $3^1/_6$. Therefore, 19/6 is the same as $3^1/_6$.

(b) $3\frac{2}{5}+2\frac{3}{10}$

We can begin by converting both of these mixed numbers into improper fractions. This would change the problem into 17/5 + 23/10. Our test fraction in this example is 5/10. Reducing this fraction, we have 5/10 = 1/2. We now know that we can multiply the top and bottom of the fraction containing a 10 in the denominator by 1. Likewise, we can multiply the top and bottom of the frac-

tion containing a 5 in the denominator by 2.

$$\frac{17}{5}+\frac{23}{10}=\frac{17\times 2}{5\times 2}+\frac{23\times 1}{10\times 1}$$

$$=\frac{34}{10}+\frac{23}{10}$$

Now that we have matching denominators, we can add the numerators. This gives us the following:

$$\frac{34}{10}+\frac{23}{10}=\frac{34+23}{10}=\frac{57}{10}$$

Again, in this example, our answer is an improper fraction. We know this since the top of our fraction is larger than the bottom. If we subtract the denominator from the numerator (57 − 10 = 47), we see that 57/10 equals $1^{47}/_{10}$. We can continue this process a few more times and get $5^7/_{10}$. Therefore, 57/10 is the same as $5^7/_{10}$.

(c) $3\frac{3}{4}+1\frac{5}{6}$

We can begin by converting both of these mixed numbers into improper fractions. This would change the problem into 15/4 + 11/6. Our test fraction in this example is 4/6. Reducing this fraction, we have 4/6 = 2/3. We now know that we can multiply the top and bottom of the fraction containing a 6 in the denominator by 2. Likewise, we can multiply the top and bottom of the fraction containing a 4 in the denominator by 3.

$$\frac{15}{4} + \frac{11}{6} = \frac{15 \times 3}{4 \times 3} + \frac{11 \times 2}{6 \times 2}$$

$$= \frac{45}{6} + \frac{22}{6}$$

Now that we have matching denominators, we can add the numerators. This gives us the following:

$$\frac{45}{6} + \frac{22}{6} = \frac{45 + 22}{6} = \frac{67}{6}$$

Again, in this example, our answer is an improper fraction. We know this since the top of our fraction is larger than the bottom. If we divide 67 by 12 we get $5^{7}/_{12}$ as our answer.

(d) $\quad 5\frac{2}{7} + 3\frac{1}{21}$

We can begin by converting both of these mixed numbers into improper fractions. This would change the problem into 37/7 + 64/21. Our test fraction in this example is 7/21. Reducing this fraction, we have 7/21 = 1/3. We now know that we can multiply the top and bottom of the fraction containing a 21 in the denominator by 1. Likewise, we can multiply the top and bottom of the fraction containing a 7 in the denominator by a 3.

$$\frac{37}{7} + \frac{64}{21} = \frac{37 \times 3}{7 \times 3} + \frac{64 \times 1}{21 \times 1}$$

$$= \frac{111}{21} + \frac{64}{21}$$

Now that we have matching denominators, we can add the numerators. This gives us the following:

$$\frac{111}{21} + \frac{64}{21} = \frac{111 + 64}{21} = \frac{175}{21}$$

Again, in this example, our answer is an improper fraction. We know this since the top of our fraction is larger than the bottom. If we subtract the denominator from the numerator (175 − 21 = 154), we see that 175/21 equals $1^{154}/_{21}$. We can continue this process a few more times and get $8^{7}/_{21}$. Therefore, 175/21 is the same as $8^{7}/_{21}$. Even though this answer is no longer an improper fraction, it is still not our final answer. If we look at the fractional part of the above answer, we see that 7/21 can be reduced. Therefore, our final answer is $8^{1}/_{3}$.

We now want to consider some unusual cases. In the examples below, we will look at problems involving fractions and whole numbers as well as problems with more than two fractions. We will see that in both cases our work is very similar to what we have already done.

Example 5

Add the following numbers.

(a) $\quad \frac{1}{2} + 3$

When adding fractions, we know that the numerators must match. In this example, the second number is not a fraction. In order to overcome this situation, we can rewrite this number as a fraction by placing it over one (as we did in the previous section). Therefore, we get 1/2 + 3/1. Now, we need to get the bottoms of these fractions to be the same value. In this case, the LCD is 2. Since the first fraction already has 2 in its denominator, we will leave it alone. In the second fraction, we have to multiply the denominator by 2 to get our LCD to appear. To keep this fraction equal, we must also multiply the numerator by the same value (2).

$$\frac{1}{2} + \frac{3}{1} = \frac{1}{2} + \frac{3 \times 2}{1 \times 2} = \frac{1}{2} + \frac{6}{2}$$

Since the denominators are both 2, we can now add the fractions.

$$\frac{1}{2} + \frac{6}{2} = \frac{1+6}{2} = \frac{7}{2}$$

This answer is an improper fraction, and we can simplify it to get $3\frac{1}{2}$.

(b) $2 + 3\frac{4}{5}$

As we did before, we want to convert both of these numbers into fractions. The first number (3) can be converted by placing it over 1. The second number (the mixed number) can be converted into an improper fraction. Therefore, we can now consider the problem as 2/1 + 19/5. The LCD in this example would be 5. In order to get a 5 to appear in the denominator of the first fraction, we need to multiply the 1 by 5. To keep the fraction equal, we must also multiply the numerator (2) by 5. Since the second fraction already contains a 5, we do not have to do anything. Therefore,

$$\frac{2}{1} + \frac{19}{5} = \frac{2 \times 5}{1 \times 5} + \frac{19}{5} = \frac{10}{5} + \frac{19}{5}$$

We are now able to add these fractions together.

$$\frac{10}{5} + \frac{19}{5} = \frac{10+19}{5} = \frac{29}{5}$$

Since our answer is an improper fraction, we know that we can convert it into a mixed number. Therefore, another way of saying 29/5 is $5\frac{4}{5}$.

(c) $\frac{1}{2} + \frac{3}{4} + \frac{7}{8}$

As with adding two fractions, in order to add three fractions, we also need a common denominator throughout the problem. It is always a good idea to use the LCD whenever possible. This keeps our numbers somewhat small. Our LCD in this example is 8. In order to get an 8 in the denominator of the first fraction, we must multiply both the top and bottom of that fraction by 4. To get an 8 to appear in the denominator of the second fraction, we must multiply the top and bottom of that fraction by 2. Since the third fraction already has 8 appearing in the denominator, we do not need to do anything with this fraction. In the first and second fractions, the numerator is also multiplied by the same value we used in the denominator so that the value of the fractions stays the same. Therefore,

$$\frac{1}{2} + \frac{3}{4} + \frac{7}{8} = \frac{1 \times 4}{2 \times 4} + \frac{3 \times 2}{4 \times 2} + \frac{7}{8}$$

We can now simplify these fractions and add them together.

$$\frac{4}{8} + \frac{6}{8} + \frac{7}{8} = \frac{4+6+7}{8} = \frac{17}{8}$$

Since the top of the fraction is larger than the bottom, we know that we have an improper fraction. If we subtract the denominator from the numerator, we can see that 17/8 is 2¹⁄₈.

(d) $\dfrac{2}{3} + \dfrac{1}{6} + \dfrac{4}{9}$

As with adding two fractions, in order to add three fractions we also need a common denominator throughout the problem. It is always a good idea to use the LCD whenever possible. This keeps our numbers somewhat small. Our LCD in this example is 18. In order to get an 18 to appear in the denominator of our first fraction, we must multiply the 3 by 6. To keep this fraction equal to the same value, we must also multiply the numerator by 6. With the second fraction, we must multiply the 6 by 3 to get an 18 to appear in the denominator. Also, we must multiply the numerator of the second fraction by 3. In the third fraction, we must multiply the denominator and numerator by 2 to keep the fraction equal and have an 18 appear in the denominator.

$$\frac{2}{3} + \frac{1}{6} + \frac{4}{9} = \frac{2 \times 6}{3 \times 6} + \frac{1 \times 3}{6 \times 3} + \frac{4 \times 2}{9 \times 2}$$

We can now simplify this and add these fractions.

$$\frac{12}{18} + \frac{3}{18} + \frac{8}{18} = \frac{12 + 3 + 8}{18} = \frac{23}{18}$$

Since this answer is an improper fraction, we know that we can subtract the denominator from the numerator and convert this into a mixed number. Therefore, 23/18 is the same as 1⁵⁄₁₈.

Questions

Add the following fractions.

1. $\dfrac{15}{31} + \dfrac{9}{31}$

2. $\dfrac{5}{7} + \dfrac{6}{7}$

3. $\dfrac{1}{2} + \dfrac{5}{16}$

4. $\dfrac{1}{6} + \dfrac{1}{9}$

5. $\dfrac{5}{7} + \dfrac{13}{14}$

Compare the following pairs of fractions. Write the appropriate symbol (=, >, <) between the two numbers.

6. $\dfrac{1}{2}$ —— $\dfrac{1}{3}$

7. $\dfrac{2}{3}$ —— $\dfrac{5}{7}$

8. $\dfrac{3}{5}$ —— $\dfrac{3}{4}$

9. $\dfrac{11}{9}$ —— $\dfrac{10}{7}$

10. $\dfrac{1}{3}$ —— $\dfrac{7}{21}$

Add the following numbers.

11. $1\dfrac{1}{3} + 5\dfrac{1}{6}$

12. $6\dfrac{4}{15} + 3\dfrac{1}{60}$

13. $\dfrac{4}{7} + 6$

14. $\dfrac{3}{4} + \dfrac{1}{8} + \dfrac{7}{16}$

15. $\dfrac{5}{7} + 9 + 1\dfrac{1}{14}$

Answers

1. $\dfrac{24}{31}$

$$\dfrac{15}{31} + \dfrac{9}{31} = \dfrac{24}{31}$$

2. $1\dfrac{4}{7}$

$$\dfrac{5}{7} + \dfrac{6}{7} = \dfrac{11}{7} \text{ or } 1\dfrac{4}{7}$$

3. $\dfrac{13}{16}$

$$\dfrac{1}{2}\dfrac{(8)}{(8)} = \dfrac{8}{16} + \dfrac{5}{16} = \dfrac{13}{16}$$

4. $\dfrac{5}{18}$

$$\dfrac{1}{6}\dfrac{(3)}{(3)} = \dfrac{3}{18}, \quad \dfrac{1}{9}\dfrac{(2)}{(2)} = \dfrac{2}{18}$$

$$\dfrac{3}{18} + \dfrac{2}{18} = \dfrac{5}{18}$$

5. $1\dfrac{9}{14}$

$$\dfrac{5}{7}\dfrac{(2)}{(2)} = \dfrac{10}{14} + \dfrac{13}{14} = \dfrac{23}{14} \text{ or } 1\dfrac{9}{14}$$

6. $>$

$$\dfrac{1}{2} = \dfrac{3}{6}, \quad \dfrac{1}{3} = \dfrac{2}{6}.$$

Since $\dfrac{3}{6} > \dfrac{2}{6}, \dfrac{1}{2} > \dfrac{1}{3}$

7. $<$

$$\dfrac{2}{3} = \dfrac{14}{21}, \quad \dfrac{5}{7} = \dfrac{15}{21}.$$

Since $\dfrac{14}{21} < \dfrac{15}{21}, \dfrac{2}{3} < \dfrac{5}{7}$

8. $<$

$$\frac{3}{5} = \frac{12}{20}, \qquad \frac{3}{4} = \frac{15}{20}.$$

Since $\frac{12}{20} < \frac{15}{20}$, $\frac{3}{5} < \frac{3}{4}$

9. $<$

$$\frac{11}{9} = \frac{77}{63}, \qquad \frac{10}{7} = \frac{90}{63}.$$

Since $\frac{77}{63} < \frac{90}{63}$, $\frac{11}{9} < \frac{10}{7}$

10. $=$

$$\frac{1}{3} = \frac{7}{21}, \text{ since } 1*21 = 3*7$$

11. $6\frac{1}{2}$

$$1\frac{1}{3} + 5\frac{1}{6} = 1\frac{2}{6} + 5\frac{1}{6}$$

$$= 6\frac{3}{6} = 6\frac{1}{2}$$

12. $9\frac{17}{60}$

$$6\frac{4}{15} + 3\frac{1}{60} = 6\frac{16}{60} + 3\frac{1}{60}$$

$$= 9\frac{17}{60}$$

13. $6\frac{4}{7}$

$$\frac{4}{7} + 6 = 6\frac{4}{7}$$

Just combine the whole number with the proper fraction.

14. $1\frac{5}{16}$

$$\frac{3}{4} + \frac{1}{8} + \frac{7}{16} = \frac{12}{16} + \frac{2}{16} + \frac{7}{16}$$

$$= \frac{21}{16} = 1\frac{5}{16}$$

15. $10\frac{11}{14}$

$$\frac{5}{7} + 9 + 1\frac{1}{14} = 10 + \frac{5}{7} + \frac{1}{14}$$

$$= 10 + \frac{10}{14} + \frac{1}{14}$$

$$= 10 + \frac{11}{14}$$

$$= 10\frac{11}{14}$$

SUBTRACTION OF FRACTIONS

As we saw before, addition and subtraction are not that much different. In order to subtract fractions, we need to have denominators that are the same. Our test fraction idea that we used when adding fractions will also be used here.

Example 1

Subtract the following numbers.

(a) $\frac{9}{13} - \frac{7}{13}$

Since the denominators match, we can go ahead and subtract the numerators. There is no need to use the test fraction in this example.

$$\frac{9}{13} - \frac{7}{13} = \frac{9-7}{13} = \frac{2}{13}$$

Notice that this fraction cannot be reduced. Therefore, our answer is 2/13.

(b) $\frac{6}{7} - \frac{5}{7}$

The denominators are the same in this example, so we can subtract the numerators. This will give us

$$\frac{6}{7} - \frac{5}{7} = \frac{6-5}{7} = \frac{1}{7}$$

Again, this fraction cannot be reduced. Thus, our answer is 1/7.

(c) $\frac{21}{25} - \frac{16}{25}$

We can subtract the numerators right away since the denominators are already the same value. This gives

$$\frac{21}{25} - \frac{16}{25} = \frac{21-16}{25} = \frac{5}{25}$$

If we look closely at our answer, we see that it can be reduced.

$$\frac{5}{25} \longrightarrow \begin{array}{c} 5/5 = 1 \\ 25/5 = 5 \end{array} \longrightarrow \frac{1}{5}$$

Therefore, our final answer is 1/5.

(d) $\frac{8}{12} - \frac{5}{12}$

As with the other examples in this section, the denominators of these fractions are already the same. This

being the case, we can subtract the numerators.

$$\frac{8}{12} - \frac{5}{12} = \frac{8-5}{12} = \frac{3}{12}$$

As with the last example, this answer can also be reduced.

$$\frac{3}{12} \longrightarrow \begin{array}{c} 3/3 = 1 \\ 12/3 = 4 \end{array} \longrightarrow \frac{1}{4}$$

Therefore, our final answer is 1/4.

Example 2

Subtract the following fractions.

(a) $\frac{5}{6} - \frac{1}{8}$

The first thing we need to do here is to get the denominators to have the same value. We can create our test fraction from the denominators in the original problem. This is 6/8 in this case. We can reduce this fraction to get 3/4. Since 3 corresponds to 6, we multiply the top and bottom of the fraction containing the 8 in the denominator by 3. Since 4 corresponds to 8, we multiply the top and bottom of the fraction containing the 6 in the denominator by 4.

$$\frac{5}{6} - \frac{1}{8} = \frac{5 \times 4}{6 \times 4} - \frac{1 \times 3}{8 \times 3}$$

If we simplify these fractions, we can subtract the numerators.

$$\frac{20}{24} - \frac{3}{24} = \frac{20-3}{24} = \frac{17}{24}$$

Since this fraction cannot be reduced, our final answer is 17/24.

(b) $\dfrac{13}{20} - \dfrac{5}{30}$

We first need to use our test fraction idea to change the denominators to the same value. In this case, our test fraction would be 20/30 or 2/3 if we reduce it. Since the 2 corresponds to the 20, we multiply the top and bottom of the fraction containing 30 by 2. Since the 3 corresponds to the 30, we multiply the top and bottom of the fraction containing the 20 by 3.

$$\dfrac{13}{20} - \dfrac{5}{30} = \dfrac{13 \times 3}{20 \times 3} - \dfrac{5 \times 2}{30 \times 2}$$

If we simplify these fractions, we can subtract the numerators.

$$\dfrac{39}{60} - \dfrac{10}{60} = \dfrac{39 - 10}{60} = \dfrac{29}{60}$$

Since this fraction cannot be reduced, our final answer is 29/60.

(c) $\dfrac{7}{12} - \dfrac{5}{9}$

First we need to use our test fraction idea to change the denominators to the same value. In this case, our test fraction would be 12/9 or 4/3 if we reduce it. Since the 4 corresponds to the 12, we multiply the top and bottom of the fraction containing the 9 by 4. Since the 3 corresponds to the 9, we multiply the top and bottom of the fraction containing 12 by 3.

$$\dfrac{7}{12} - \dfrac{5}{9} = \dfrac{7 \times 3}{12 \times 3} - \dfrac{5 \times 4}{9 \times 4}$$

Next, we can simplify these fractions and then subtract the numerators. This gives us

$$\dfrac{21}{36} - \dfrac{20}{36} = \dfrac{21 - 20}{36} = \dfrac{1}{36}$$

Since this fraction cannot be reduced, our final answer is 1/36.

(d) $\dfrac{13}{20} - \dfrac{5}{8}$

We first need to use our test fraction idea to change the denominators to the same value. In this case, our test fraction would be 20/8 or 5/2 if we reduce it. Since the 5 corresponds to the 20, we multiply the top and bottom of the fraction containing 8 by 5. Since the 2 corresponds to the 8, we multiply the top and bottom of the fraction containing the 20 by 2.

$$\dfrac{13}{20} - \dfrac{5}{8} = \dfrac{13 \times 2}{20 \times 2} - \dfrac{5 \times 5}{8 \times 5}$$

Now we can simplify these fractions and subtract the numerators.

$$\dfrac{26}{40} - \dfrac{25}{40} = \dfrac{26 - 25}{40} = \dfrac{1}{40}$$

Since this fraction cannot be reduced, our final answer is 1/40.

Example 3

Subtract the following numbers.

(a) $3\dfrac{1}{2} - 2\dfrac{3}{4}$

When dealing with mixed numbers, the first step that we want to do is convert everything into improper fractions. In this example, our problem becomes 7/2 − 11/4. Now we need to get the denominators to have the same value. This can be done using our test fraction idea. Our test fraction is 2/4, or 1/2 if we reduce it. Since the 1 corresponds to the 2, we multiply the top and bottom of the fraction that contains 4 by 1. Since the 2 corresponds to the 4, we multiply the top and bottom of the fraction containing 2 by 2. Therefore,

$$\frac{7}{2} - \frac{11}{4} = \frac{7 \times 2}{2 \times 2} - \frac{11 \times 1}{4 \times 1}$$

If we simplify these fractions, we can then subtract the numerators. Thus,

$$\frac{14}{4} - \frac{11}{4} - \frac{14 - 11}{4} = \frac{3}{4}$$

Since this fraction cannot be reduced, our final answer is 3/4.

(b) $\quad 2\dfrac{1}{6} - 1\dfrac{7}{8}$

When dealing with mixed numbers, the first step that we want to do is convert everything into improper fractions. In this example, our problem becomes 13/6 – 15/8. Now we need to get the denominators to have the same value. This can be done using our test fraction idea. Our test fraction is 6/8, or 3/4 if we reduce it. Since the 3 corresponds to the 6, we multiply the top and bottom of the fraction that contains 8 by 3. Since the 4 corresponds to the 8 we multiply the top and bottom of the fraction containing 6 by 4. Therefore,

$$\frac{13}{6} - \frac{15}{8} = \frac{13 \times 4}{6 \times 4} - \frac{15 \times 3}{8 \times 3}$$

If we simplify these fractions, we can then subtract the numerators. Thus,

$$\frac{52}{24} - \frac{45}{24} = \frac{52 - 45}{24} = \frac{7}{24}$$

Since this fraction cannot be reduced, our final answer is 7/24.

(c) $\quad 3\dfrac{1}{14} - 2\dfrac{5}{7}$

When dealing with mixed numbers, the first step that we want to do is convert everything into improper fractions. In this example, our problem becomes 43/14 – 19/7. Now we need to get the denominators to have the same value. This can be done using our test fraction idea. Our test fraction is 14/7, or 2/1 if we reduce it. Since the 2 corresponds to the 14, we multiply the top and bottom of the fraction that contains 7 by 2. Since the 1 corresponds to the 7, we multiply the top and bottom of the fraction containing 14 by 1. Therefore,

$$\frac{43}{14} - \frac{19}{7} = \frac{43 \times 1}{14 \times 1} - \frac{19 \times 2}{7 \times 2}$$

If we simplify these fractions, we can then subtract the numerators. Thus,

$$\frac{43}{14} - \frac{38}{14} = \frac{43 - 38}{14} = \frac{5}{14}$$

Since this fraction cannot be reduced, our final answer is 5/14.

(d) $\quad 4\dfrac{1}{6} - 1\dfrac{1}{9}$

When dealing with mixed numbers, the first step that we want to do is convert everything into improper fractions. In this example, our problem becomes 25/6 – 10/9. Now we need to get the denominators to have the same value. This can be done using our test fraction idea. Our test fraction is 6/9, or 2/3 if we reduce it. Since the 2 corresponds to the 6, we multiply the top and bottom of the fraction that contains 9 by 2. Since the 3 corresponds to the 9, we multiply the top and bottom of the fraction containing 6 by 3. Therefore,

$$\frac{25}{6} - \frac{10}{9} = \frac{25 \times 3}{6 \times 3} - \frac{10 \times 2}{9 \times 2}$$

If we simplify these fractions, we can then subtract the numerators. Thus,

$$\frac{75}{18} - \frac{20}{18} = \frac{75 - 20}{18} = \frac{55}{18}$$

Since the numerator of this fraction is larger than the denominator, we know that we have an improper fraction that we can rewrite as a mixed number. If we subtract the denominator from the numerator, we see that $1^{37}/_{18}$ is the same as 55/18. We can continue this and see that $3^1/_{18}$ equals 55/18. Therefore, our final answer is $3^1/_{18}$.

Questions

Subtract the following numbers.

1. $\dfrac{2}{3} - \dfrac{1}{3}$

2. $\dfrac{3}{4} - \dfrac{1}{4}$

3. $\dfrac{9}{10} - \dfrac{3}{10}$

4. $\dfrac{25}{37} - \dfrac{10}{37}$

5. $\dfrac{1}{3} - \dfrac{1}{4}$

6. $\dfrac{11}{36} - \dfrac{5}{24}$

7. $\dfrac{2}{3} - \dfrac{26}{51}$

8. $2\dfrac{1}{3} - 1\dfrac{7}{9}$

9. $5\dfrac{1}{2} - 2\dfrac{5}{16}$

10. $7\dfrac{19}{36} - 5\dfrac{11}{24}$

Answers

1. $\dfrac{1}{3}$

$$\frac{2}{3} - \frac{1}{3} = \frac{1}{3}$$

Just subtract in the numerator.

2. $\dfrac{1}{2}$

$$\frac{3}{4} - \frac{1}{4} = \frac{2}{4} = \frac{1}{2}$$

Subtract in the numerator. Reduce to lowest terms.

3. $\dfrac{3}{5}$

$$\frac{9}{10} - \frac{3}{10} = \frac{6}{10} = \frac{3}{5}$$

Subtract in the numerator. Reduce to lowest terms.

4. $\dfrac{15}{37}$

$$\dfrac{25}{37} - \dfrac{10}{27} = \dfrac{15}{27}$$

Just subtract in the numerator.

5. $\dfrac{1}{12}$

$$\dfrac{1}{3} - \dfrac{1}{4} = \dfrac{4}{12} - \dfrac{3}{12} = \dfrac{1}{12}$$

Convert each fraction to twelfths, where 12 is the least common denominator of 3, 4.

6. $\dfrac{7}{72}$

$$\dfrac{11}{36} - \dfrac{5}{24} = \dfrac{22}{72} - \dfrac{15}{72} = \dfrac{7}{72}$$

Convert each fraction to a denominator of 72, which is the least common denominator of 24, 36.

7. $\dfrac{8}{51}$

$$\dfrac{2}{3} - \dfrac{26}{51} = \dfrac{34}{51} - \dfrac{26}{51} = \dfrac{8}{51}$$

Convert 2/3 to a fraction with a denominator of 51. Subtract in the numerator.

8. $\dfrac{5}{9}$

$$2\dfrac{1}{3} - 1\dfrac{7}{9} = 2\dfrac{3}{9} - 1\dfrac{7}{9}$$
$$= 1\dfrac{12}{9} - 1\dfrac{7}{9} = \dfrac{5}{9}$$

Convert first fractional part to ninths. Change 2³/₉ to 1¹²/₉. Subtract whole parts and fractional parts separately.

9. $3\dfrac{3}{16}$

$$5\dfrac{1}{2} - 2\dfrac{5}{16} = 5\dfrac{8}{16} - 2\dfrac{5}{16} = 3\dfrac{3}{16}$$

Convert first fractional part to sixteenths. Subtract whole parts and fractional parts separately.

10. $2\dfrac{5}{72}$

$$7\dfrac{19}{36} - 5\dfrac{11}{24} = 7\dfrac{38}{72} - 5\dfrac{33}{72} = 2\dfrac{5}{72}$$

Convert both fractional parts to a denominator of 72, which is the least common denominator of 24, 36. Subtract whole parts and fractional parts separately.

MULTIPLICATION OF FRACTIONS

Now that we understand adding and subtracting fractions, we want to switch gears and discuss multiplying fractions. Even though this topic may sound more involved, it is actually easier. In multiplication, there is no need for a Lowest Common Denominator (LCD). Multiplying fractions is one of the most intuitive operations. For the most part, we just multiply straight across. We multiply the numerators together to get the resulting "answer" numerator, and then multiply the denominators together to get the resulting "answer" denominator. We can then reduce this "answer" (if possible) to get our final answer. As we move through this section, we see a few "short-cuts" that can be used to make the problems more tolerable. Initially, we will just multiply straight across.

Example 1

Multiply the following numbers.

(a) $\dfrac{1}{2} \times \dfrac{1}{3}$

When multiplying fractions, we just multiply straight across. We multiply the numerators and then the denominators. In this example, we get $1 \times 1 = 1$ in the numerator and $2 \times 3 = 6$ in the denominator.

$$\frac{1}{2} \times \frac{1}{3} = \frac{1 \times 1}{2 \times 3} = \frac{1}{6}$$

Since 1/6 cannot be reduced, our final answer is 1/6.

(b) $\dfrac{3}{4} \times \dfrac{5}{7}$

When multiplying fractions, we just multiply straight across. We multiply the numerators and then the denominators. In this example, we get $3 \times 5 = 15$ in the numerator and $4 \times 7 = 28$ in the denominator.

$$\frac{3}{4} \times \frac{5}{7} = \frac{3 \times 5}{4 \times 7} = \frac{15}{28}$$

Since 15/28 cannot be reduced, our final answer is 15/28.

(c) $\dfrac{2}{5} \times \dfrac{4}{9}$

When multiplying fractions, we just multiply straight across. We multiply the numerators and then the denominators. In this example, we get $2 \times 4 = 8$ in the numerator and $5 \times 9 = 45$ in the denominator.

$$\frac{2}{5} \times \frac{4}{9} = \frac{2 \times 4}{5 \times 9} = \frac{8}{45}$$

Since 8/45 cannot be reduced, our final answer is 8/45.

(d) $\dfrac{4}{7} \times \dfrac{1}{2}$

When multiplying fractions, we just multiply straight across. We multiply the numerators and then the denominators. In this example, we get $4 \times 1 = 4$ in the numerator and $7 \times 2 = 14$ in the denominator.

$$\frac{4}{7} \times \frac{1}{2} = \frac{4 \times 1}{7 \times 2} = \frac{4}{14}$$

Notice that this fraction can be reduced. Both the top and the bottom are divisible by 2.

$$\frac{4}{14} \begin{array}{c} \rightarrow 4/2 = 2 \rightarrow \\ \rightarrow 14/2 = 7 \rightarrow \end{array} \frac{2}{7}$$

Therefore, our final answer is 2/7.

Even though the method described above will always work, there is a "short-cut" that can help us out. A good example of these special problems is example 1(d) above. Instead of waiting and reducing at the end of the problem, in some problems we are able to reduce in the beginning and work with smaller numbers. In order to this, we must have a common factor appearing in both a numerator and a denominator. The difference here is that they do not have to be in the same fraction. As long as one is above the fraction bar, one is below the fraction bar, and the operation between the fractions is multiplication, we can do a type of cancellation. Let us reconsider the above problem.

$$\frac{4}{7} \times \frac{1}{2}$$

Notice that a common factor of 2 appears in the numerator of the first fraction and in the denominator of the second fraction. Two can be divided evenly into both 4 and 2. If we do this cancellation, we replace the 4 with a 2 (4/2) and replace the 2 with a 1 (2/2).

$$\frac{4}{7} \times \frac{1}{2} = \frac{\overset{2}{\cancel{4}}}{7} \times \frac{1}{\underset{1}{\cancel{2}}} = \frac{2}{7} \times \frac{1}{1}$$

After this cancellation has been done, we multiply the fractions together as we talked about earlier.

$$\frac{2}{7} \times \frac{1}{1} = \frac{2 \times 1}{7 \times 1} = \frac{2}{7}$$

Since 2/7 cannot be reduced, our answer this time is 2/7. Notice that in both cases (here and example 1(d)), we get the same value for our answer.

Example 2

Multiply the following numbers. Cancel where possible.

(a) $\frac{4}{5} \times \frac{3}{8}$

Instead of multiplying straight across, we want to consider if cancellation is possible. In this example, notice the top of the first fraction and the bottom of the second fraction. Four can be divided into both 4 and 8 evenly. Therefore, we get

$$\frac{4}{5} \times \frac{3}{8} = \frac{\overset{1}{\cancel{4}}}{5} \times \frac{3}{\underset{2}{\cancel{8}}} = \frac{1}{5} \times \frac{3}{2}$$

Now that we have canceled where possible, we can multiply the fractions straight across.

$$\frac{1}{5} \times \frac{3}{2} = \frac{1 \times 3}{5 \times 2} = \frac{3}{10}$$

Since 3/10 cannot be reduced, our final answer is 3/10. Notice in this example, by canceling first, the numbers we multiply together are smaller and easier to deal with.

(b) $\frac{5}{6} \times \frac{12}{13}$

Instead of multiplying straight across, we want to consider if cancellation is possible. In this example, notice the bottom of the first fraction and the top of the second fraction. Six can be divided into both 6 and 12 evenly. Therefore, we get

$$\frac{5}{6} \times \frac{12}{13} = \frac{5}{\underset{1}{\cancel{6}}} \times \frac{\overset{2}{\cancel{12}}}{13} = \frac{5}{1} \times \frac{2}{13}$$

Now that we have canceled where possible, we can multiply the fractions straight across.

$$\frac{5}{1} \times \frac{2}{13} = \frac{5 \times 2}{1 \times 13} = \frac{10}{13}$$

Since 10/13 cannot be reduced, our final answer is 10/13.

(c) $\dfrac{4}{7} \times \dfrac{14}{15}$

Instead of multiplying straight across, we want to consider if cancellation is possible. In this example, notice the bottom of the first fraction and the top of the second fraction. Seven can be divided into both 7 and 14 evenly. Therefore, we get

$$\dfrac{4}{7} \times \dfrac{14}{15} = \dfrac{4}{\overset{}{\underset{1}{\cancel{7}}}} \times \dfrac{\overset{2}{\cancel{14}}}{15} = \dfrac{4}{1} \times \dfrac{2}{15}$$

Now that we have canceled where possible, we can multiply the fractions straight across.

$$\dfrac{4}{1} \times \dfrac{2}{15} = \dfrac{4 \times 2}{1 \times 15} = \dfrac{8}{15}$$

Since 8/15 cannot be reduced, our final answer is 8/15.

(d) $\dfrac{2}{3} \times \dfrac{9}{16}$

Instead of multiplying straight across, we want to consider if cancellation is possible. In this example, notice the top of the first fraction and the bottom of the second fraction. Two can be divided into both 2 and 16 evenly. Also, consider the bottom of the first fraction and the top of the second fraction. Three can be divided into both 3 and 9 evenly. Therefore, we get

$$\dfrac{2}{3} \times \dfrac{9}{16} = \dfrac{\overset{1}{\cancel{2}}}{\underset{1}{\cancel{3}}} \times \dfrac{\overset{3}{\cancel{9}}}{\underset{8}{\cancel{16}}} = \dfrac{1}{1} \times \dfrac{3}{8}$$

Now that we have canceled where possible, we can multiply the fractions straight across.

$$\dfrac{1}{1} \times \dfrac{3}{8} = \dfrac{1 \times 3}{1 \times 8} = \dfrac{3}{8}$$

Since 3/8 cannot be reduced, our final answer is 3/8.

Now that we understand how to multiply fractions, we want to move our discussion to mixed numbers. As we did in the addition and subtraction sections, we will begin by rewriting the mixed number as an improper fraction. Then we can cancel where possible, and finally multiply straight across.

Example 3

Multiply the following numbers.

(a) $1\dfrac{1}{2} \times 1\dfrac{1}{3}$

We begin by converting both of these mixed numbers into improper fractions. In this example, we have 3/2 × 4/3. Instead of multiplying straight across, we want to consider if cancellation is possible. In this example, notice the top of the first fraction and the bottom of the second fraction. Three can be divided into both 3 and 3 evenly. Also, consider the bottom of the first fraction and the top of the second fraction. Two can be divided into both 2 and 4 evenly. Therefore, we get

$$\dfrac{3}{2} \times \dfrac{4}{3} = \dfrac{\overset{1}{\cancel{3}}}{\underset{1}{\cancel{2}}} \times \dfrac{\overset{2}{\cancel{4}}}{\underset{1}{\cancel{3}}} = \dfrac{1}{1} \times \dfrac{2}{1}$$

Now that we have canceled where

possible, we can multiply the fractions straight across.

$$\frac{1}{1} \times \frac{2}{1} = \frac{1 \times 2}{1 \times 1} = \frac{2}{1}$$

Since we can rewrite 2/1 as 2, our final answer is 2.

(b) $3\frac{3}{4} \times 2\frac{2}{5}$

We begin by converting both of these mixed numbers into improper fractions. In this example, we have 15/4 × 12/5. Instead of multiplying straight across, we want to consider if cancellation is possible. In this example, notice the top of the first fraction and the bottom of the second fraction. Five can be divided into both 15 and 5 evenly. Also, consider the bottom of the first fraction and the top of the second fraction. Four can be divided into both 4 and 12 evenly. Therefore, we get

$$\frac{15}{4} \times \frac{12}{5} = \frac{\overset{3}{\cancel{15}}}{\underset{1}{\cancel{4}}} \times \frac{\overset{3}{\cancel{12}}}{\underset{1}{\cancel{5}}} = \frac{3}{1} \times \frac{3}{1}$$

Now that we have canceled where possible, we can multiply the fractions straight across.

$$\frac{3}{1} \times \frac{3}{1} = \frac{3 \times 3}{1 \times 1} = \frac{9}{1}$$

Since we can rewrite 9/1 as 9, our final answer is 9.

(c) $5\frac{6}{7} \times 4\frac{2}{3}$

We begin by converting both of these mixed numbers into improper frac-

tions. In this example, we have 41/7 × 14/3. Instead of multiplying straight across, we want to consider if cancellation is possible. In this example, notice the bottom of the first fraction and the top of the second fraction. Seven can be divided into both 7 and 14 evenly. Therefore, we get

$$\frac{41}{7} \times \frac{14}{3} = \frac{41}{\underset{1}{\cancel{7}}} \times \frac{\overset{2}{\cancel{14}}}{3} = \frac{41}{1} \times \frac{2}{3}$$

Now that we have canceled where possible, we can multiply the fractions straight across.

$$\frac{41}{1} \times \frac{2}{3} = \frac{41 \times 2}{1 \times 3} = \frac{82}{3}$$

Since the numerator is larger than the denominator, we know that we are dealing with a fraction larger than 1. If we subtract the denominator from the numerator, we see that 82/3 is the same as 1⁷⁹/₃. If we continued this process, we would get 27¹/₃. Therefore, our final answer is 27¹/₃. We could also divide 82 by 3, in which case we would obtain 27 with a remainder of 1. As shown this translates to 27¹/₃.

(d) $4\frac{1}{2} \times 16$

We begin by converting both of these mixed numbers into improper fractions. In this example, we have 9/2 × 16/1. Instead of multiplying straight across, we want to consider if cancellation is possible. In this example, notice the bottom of the first fraction and the top of the second fraction. Two can be divided into both 2 and 16 evenly. Therefore, we get

$$\frac{9}{2} \times \frac{16}{1} = \frac{9}{\cancel{2}} \times \frac{\cancel{16}\,^{8}}{1} = \frac{9}{1} \times \frac{8}{1}$$

Now that we have canceled where possible, we can multiply the fractions together straight across.

$$\frac{9}{1} \times \frac{8}{1} = \frac{9 \times 8}{1 \times 1} = \frac{72}{1}$$

Since we can rewrite 72/1 as 72, our final answer is 72.

Questions

Multiply the following numbers.

1. $\dfrac{1}{5} \times \dfrac{7}{2}$

2. $\dfrac{3}{4} \times \dfrac{5}{7}$

3. $\dfrac{7}{13} \times \dfrac{3}{14}$

4. $\dfrac{1}{5} \times \dfrac{10}{13}$

5. $\dfrac{4}{5} \times \dfrac{35}{36}$

6. $\dfrac{10}{27} \times \dfrac{3}{20}$

7. $1\dfrac{1}{5} \times 3\dfrac{1}{2}$

8. $4\dfrac{2}{7} \times 5\dfrac{1}{6}$

9. $7\dfrac{1}{2} \times 1\dfrac{1}{3}$

10. $3\dfrac{4}{5} \times 2\dfrac{9}{38}$

Answers

1. $\dfrac{7}{10}$

$$\frac{1}{5} \times \frac{7}{2} = \frac{7}{10}$$

Multiply numerators, then multiply denominators.

2. $\dfrac{15}{28}$

$$\frac{3}{4} \times \frac{5}{7} = \frac{15}{28}$$

Multiply numerators, then multiply denominators.

3. $\dfrac{3}{26}$

$$\frac{7}{13} \times \frac{3}{14} = \frac{\cancel{7}\,^{1}}{13} \times \frac{3}{\cancel{14}\,_{2}} = \frac{3}{26}$$

Divide 7, 14 by 7 to produce 1, 2. Multiply numerators, then multiply denominators.

4. $\dfrac{2}{13}$

$$\frac{1}{5} \times \frac{10}{13} = \frac{1}{\underset{1}{5}} \times \frac{\overset{2}{\cancel{10}}}{13} = \frac{2}{13}$$

Divide 5, 10 by 5 to produce 1, 2. Multiply numerators, then multiply denominators.

5. $\dfrac{7}{9}$

$$\frac{4}{5} \times \frac{35}{36} = \frac{\overset{1}{\cancel{4}}}{\underset{1}{5}} \times \frac{\overset{7}{\cancel{35}}}{\underset{9}{36}} = \frac{7}{9}$$

Divide 4, 36 by 4 to produce 1, 9. Divide 5, 35 by 5 to produce 1, 7. Multiply numerators, then multiply denominators.

6. $\dfrac{1}{18}$

$$\frac{10}{27} \times \frac{3}{20} = \frac{\overset{1}{\cancel{10}}}{\underset{9}{\cancel{27}}} \times \frac{\overset{1}{\cancel{3}}}{\underset{2}{\cancel{20}}} = \frac{1}{18}$$

Divide 10, 20 by 10 to produce 1, 2. Divide 3, 27 by 3 to produce 1, 9. Multiply numerators, then multiply denominators.

7. $4\dfrac{1}{5}$

$$1\frac{1}{5} \times 3\frac{1}{2} = \frac{6}{5} \times \frac{7}{2}$$

$$= \frac{\overset{3}{\cancel{6}}}{5} \times \frac{7}{\underset{1}{2}} = \frac{21}{5} = 4\frac{1}{5}$$

Convert mixed fractions to improper fractions. Divide 2, 6, by 2 to produce 1, 3. Multiply numerators, then denominators. Convert improper fraction to mixed fraction.

8. $22\dfrac{1}{7}$

$$4\frac{2}{7} \times 5\frac{1}{6} = \frac{30}{7} \times \frac{31}{6}$$

$$= \frac{\overset{5}{\cancel{30}}}{7} \times \frac{31}{\underset{1}{6}} = \frac{155}{7} = 22\frac{1}{7}$$

Convert mixed fractions to improper fractions. Divide 6, 30 by 6 to produce 1, 5. Multiply numerators, then denominators. Convert improper fraction to mixed fraction.

9. 10

$$7\frac{1}{2} \times 1\frac{1}{3} = \frac{15}{2} \times \frac{4}{3}$$

$$= \frac{\overset{5}{\cancel{15}}}{\underset{1}{2}} \times \frac{\overset{1}{\cancel{4}}}{\underset{1}{3}} = \frac{10}{1} = 10$$

Convert mixed fractions to improper fractions. Divide 3, 15 by 3 to produce 1, 5. Divide 2, 4 by 2 to produce 1, 2. Multiply numerators, then denominators. Reduce 10/1 to 10.

10. $8\dfrac{1}{2}$

$$3\dfrac{4}{5} \times 2\dfrac{9}{38} = \dfrac{19}{5} \times \dfrac{85}{38}$$

$$= \dfrac{\cancel{19}}{\underset{1}{5}} \times \dfrac{\overset{17}{\cancel{85}}}{\underset{2}{38}} = \dfrac{17}{2} = 8\dfrac{1}{2}$$

Convert mixed fractions to improper fractions. Divide 19, 38 by 19 to produce 1, 2. Divide 5, 85 by 5 to produce 1, 17. Multiply numerators, then denominators. Convert improper fraction to mixed fraction.

DIVISION OF FRACTIONS

As with multiplication, division does not require the use of a Lowest Common Denominator (LCD). In some ways, division is similar to multiplication. In order to divide fractions, we rewrite the first fraction, flip the second fraction, and then multiply. Mathematicians love to change problems into ones that they already know how to simplify. In this case, they change division into multiplication. Instead of "flipping" the second fraction, sometimes this will be referred to as forming the **reciprocal**. Whether we refer to it as forming the reciprocal or flipping, we are just switching the positions of the numerator and denominator.

Example 1

Divide the following fractions.

(a) $\dfrac{2}{3}\Big/\dfrac{5}{7}$

In order to divide these fractions, we rewrite the first fraction, flip the second fraction, and then multiply. In this problem, we have

$$\dfrac{2 \times 7}{3 \times 5} = \dfrac{14}{15}$$

Since this fraction cannot be reduced, our final answer is 14/15. Notice that no cancellation was possible in this example. Remember that we want to re-construct our problem into multiplication before any canceling is done. Even though 5/7 is a normal fraction, after flipping, it is improper.

(b) $\dfrac{1}{2}\Big/\dfrac{3}{4}$

In order to divide these fractions, we rewrite the first fraction, flip the second fraction, and then multiply. In this problem, we have

$$\dfrac{1}{2} \times \dfrac{4}{3} = \dfrac{1}{\underset{1}{\cancel{2}}} \times \dfrac{\overset{2}{\cancel{4}}}{3} = \dfrac{1}{1} \times \dfrac{2}{3} = \dfrac{1 \times 2}{1 \times 3} = \dfrac{2}{3}$$

Since this fraction cannot be reduced, our final answer is 2/3. Notice that we canceled where possible. Remember that we want to re-construct our problem into multiplication before any canceling is done. Even though 3/4 is a normal fraction, after flipping, it is improper.

(c) $\dfrac{6}{13}\Big/\dfrac{12}{5}$

In order to divide these fractions, we rewrite the first fraction, flip the second fraction, and then multiply. In this problem, we have

$$\dfrac{6}{13}\times\dfrac{5}{12}=\dfrac{6}{13}\times\dfrac{5}{\overset{}{\underset{2}{\cancel{12}}}}\;\overset{1}{}$$

$$=\dfrac{1}{13}\times\dfrac{5}{2}=\dfrac{1\times5}{13\times2}=\dfrac{5}{26}$$

Since 5/26 cannot be reduced, our final answer is 5/26. Notice that we canceled where possible. Remember that we want to re-construct our problem into multiplication before any canceling is done.

(d) $\dfrac{5}{9}\Big/\dfrac{10}{18}$

In order to divide these fractions, we rewrite the first fraction, flip the second fraction, and then multiply. In this problem, we have

$$\dfrac{5}{9}\times\dfrac{18}{10}=\dfrac{\overset{1}{\cancel{5}}}{\underset{1}{9}}\times\dfrac{\overset{2}{\cancel{18}}}{\underset{2}{\cancel{10}}}$$

$$=\dfrac{1}{1}\times\dfrac{\overset{1}{\cancel{2}}}{\underset{1}{\cancel{2}}}=\dfrac{1\times1}{1\times1}=\dfrac{1}{1}$$

Since 1/1 can be rewritten as 1, our final answer is 1. Notice that we canceled where possible. Remember that we want to re-construct our problem into multiplication before

any canceling is done.

Example 2

Divide the following fractions.

(a) $1\dfrac{1}{2}\Big/2\dfrac{3}{4}$

As before, we begin by writing the mixed numbers as improper fractions. In this example, we get 3/2 ÷ 11/4. We now rewrite the first fraction, flip the second fraction, and then multiply.

$$\dfrac{3}{2}\times\dfrac{4}{11}=\dfrac{3}{\underset{1}{\cancel{2}}}\times\dfrac{\overset{2}{\cancel{4}}}{11}=\dfrac{3}{1}\times\dfrac{2}{11}$$

$$=\dfrac{3\times2}{1\times11}=\dfrac{6}{11}$$

Since this fraction cannot be reduced, our final answer is 6/11.

(b) $1\dfrac{1}{7}\Big/1\dfrac{3}{5}$

As before, we begin by writing the mixed numbers as improper fractions. In this example, we get 8/7 ÷ 8/5. We now rewrite the first fraction, flip the second fraction, and then multiply.

$$\dfrac{8}{7}\times\dfrac{5}{8}=\dfrac{\overset{1}{\cancel{8}}}{7}\times\dfrac{5}{\underset{1}{\cancel{8}}}=\dfrac{1}{7}\times\dfrac{5}{1}$$

$$=\dfrac{1\times5}{7\times1}=\dfrac{5}{7}$$

Since this fraction cannot be reduced, our final answer is 5/7.

(c) $2\dfrac{1}{5}\Big/1\dfrac{7}{15}$

As before, we begin by writing the mixed numbers as improper fractions. In this example, we get 11/5 ÷ 22/15. We now rewrite the first fraction, flip the second fraction, and then multiply.

$$\dfrac{11}{5}\times\dfrac{15}{22}=\dfrac{\overset{1}{\cancel{11}}}{\underset{1}{5}}\times\dfrac{\overset{3}{\cancel{15}}}{\underset{2}{\cancel{22}}}$$

$$=\dfrac{1}{1}\times\dfrac{3}{2}=\dfrac{1\times3}{1\times2}=\dfrac{3}{2}$$

Since the numerator is larger than the denominator, we know that we are dealing with a number larger than 1. By subtracting the denominator from the numerator, we see that 3/2 equals 1½. Therefore, our final answer is 1½.

(d) $4\dfrac{2}{9}\Big/3\dfrac{1}{6}$

As before, we begin by writing the mixed numbers as improper fractions. In this example, we get 38/9 ÷ 19/6. We now rewrite the first fraction, flip the second fraction, and then multiply.

$$\dfrac{38}{9}\times\dfrac{6}{19}=\dfrac{\overset{2}{38}}{\underset{3}{9}}\times\dfrac{\overset{2}{6}}{\underset{1}{\cancel{19}}}$$

$$=\dfrac{2}{3}\times\dfrac{2}{1}=\dfrac{2\times2}{3\times1}=\dfrac{4}{3}$$

Since the numerator is larger than the denominator, we know that we are

dealing with a number larger than 1. By subtracting the denominator from the numerator, we see that 4/3 equals 1⅓. Therefore, our final answer is 1⅓.

Questions

Divide the following fractions.

1. $\dfrac{1}{4}\Big/\dfrac{7}{8}$

2. $\dfrac{2}{3}\Big/\dfrac{4}{7}$

3. $\dfrac{5}{9}\Big/\dfrac{10}{13}$

4. $\dfrac{13}{5}\Big/\dfrac{26}{50}$

5. $\dfrac{8}{24}\Big/\dfrac{16}{12}$

6. $1\dfrac{3}{7}\Big/1\dfrac{1}{14}$

7. $3\dfrac{1}{5}\Big/3\dfrac{1}{8}$

8. $2\dfrac{9}{13}\Big/3\dfrac{23}{39}$

9. $5\dfrac{4}{9}\Big/1\dfrac{3}{18}$

10. $4\dfrac{4}{15}\Big/2\dfrac{2}{5}$

Answers

1. $\dfrac{2}{7}$

$$\dfrac{1}{4}\Big/\dfrac{7}{8} = \dfrac{1}{4}\times\dfrac{8}{7} = \dfrac{1}{4}\times\dfrac{\overset{2}{\cancel{8}}}{7} = \dfrac{2}{7}$$
$$\underset{1}{}$$

2. $1\dfrac{1}{6}$

$$\dfrac{2}{3}\Big/\dfrac{4}{7} = \dfrac{2}{3}\times\dfrac{7}{4} = \dfrac{\overset{1}{\cancel{2}}}{3}\times\dfrac{7}{\underset{2}{\cancel{4}}} = \dfrac{7}{6} = 1\dfrac{1}{6}$$

3. $\dfrac{13}{18}$

$$\dfrac{5}{9}\Big/\dfrac{10}{13} = \dfrac{5}{9}\times\dfrac{13}{10} = \dfrac{5}{9}\times\dfrac{13}{\underset{2}{\overset{1}{\cancel{10}}}} = \dfrac{13}{18}$$

4. 5

$$\dfrac{13}{5}\Big/\dfrac{26}{50} = \dfrac{13}{5}\times\dfrac{50}{26}$$
$$= \dfrac{\overset{1}{\cancel{13}}}{5}\times\dfrac{\overset{10}{\cancel{50}}}{\underset{2}{\cancel{26}}} = \dfrac{10}{2} = 5$$
$$\underset{1}{}$$

5. $\dfrac{1}{4}$

$$\dfrac{8}{24}\Big/\dfrac{16}{12} = \dfrac{8}{24}\times\dfrac{12}{16}$$
$$= \dfrac{8}{\underset{2}{\cancel{24}}}\times\dfrac{\overset{1}{\cancel{12}}}{\underset{2}{\cancel{16}}}^{\,1} = \dfrac{1}{4}$$

6. $1\dfrac{1}{3}$

$$1\dfrac{3}{7}\Big/1\dfrac{1}{14} = \dfrac{10}{7}\Big/\dfrac{15}{14} = \dfrac{10}{7}\times\dfrac{14}{15}$$
$$= \dfrac{\overset{2}{\cancel{10}}}{\underset{1}{\cancel{7}}}\times\dfrac{\overset{2}{\cancel{14}}}{\underset{3}{\cancel{15}}} = \dfrac{4}{3} = 1\dfrac{1}{3}$$

7. $1\dfrac{3}{125}$

$$3\dfrac{1}{5}\Big/3\dfrac{1}{8} = \dfrac{16}{5}\Big/\dfrac{25}{8} = \dfrac{16}{5}\times\dfrac{8}{25}$$
$$= \dfrac{128}{125} = 1\dfrac{3}{125}$$

8. $\dfrac{3}{4}$

$$2\dfrac{9}{13}\Big/3\dfrac{23}{39} = \dfrac{35}{13}\Big/\dfrac{140}{39} = \dfrac{35}{13}\times\dfrac{39}{140}$$
$$= \dfrac{\overset{1}{\cancel{35}}}{\underset{1}{\cancel{13}}}\times\dfrac{\overset{3}{\cancel{39}}}{\underset{4}{\cancel{140}}} = \dfrac{3}{4}$$

9. $4\dfrac{2}{3}$

10. $1\dfrac{7}{9}$

$$5\dfrac{4}{9}\Big/1\dfrac{3}{18}=\dfrac{49}{9}\Big/\dfrac{21}{18}=\dfrac{49}{9}\times\dfrac{18}{21}$$

$$=\dfrac{49}{\underset{1}{\cancel{9}}}\times\dfrac{\overset{2}{\cancel{18}}}{\underset{3}{\cancel{21}}}^{7}=\dfrac{14}{3}=4\dfrac{2}{3}$$

$$4\dfrac{4}{15}\Big/2\dfrac{2}{5}=\dfrac{64}{15}\Big/\dfrac{12}{5}=\dfrac{64}{15}\times\dfrac{5}{12}$$

$$=\dfrac{\overset{16}{\cancel{64}}}{\underset{3}{\cancel{15}}}\times\dfrac{\overset{1}{\cancel{5}}}{\underset{3}{\cancel{12}}}=\dfrac{16}{9}=1\dfrac{7}{9}$$

☞ Practice: Fractions

DIRECTIONS: Solve each problem below.

1. For the fraction $\dfrac{5}{7}$, the 7 is called the _____.

2. The reduced form of $\dfrac{6}{21}$ is _____.

3. The reduced form of $\dfrac{36}{44}$ is _____.

4. $\dfrac{9}{2}$ written as a mixed number is _____.

5. $\dfrac{28}{5}$ written as a mixed number is _____.

6. $3\dfrac{1}{3}$ written as an improper fraction is _____.

7. $2\dfrac{3}{8}$ written as an improper fraction is _____.

8. $\dfrac{5}{13}+\dfrac{4}{13}=$

9. $\dfrac{3}{5}+\dfrac{1}{2}=$

10. $\dfrac{4}{15}+\dfrac{1}{3}=$

11. $\dfrac{2}{3}$ _____ $\dfrac{8}{13}$. Fill in the blank with <, =, or >.

12. $\dfrac{12}{16}$ _____ $\dfrac{15}{20}$. Fill in the blank with <, =, or >.

13. $1\dfrac{3}{7}+2\dfrac{4}{9}=$

14. $3\dfrac{5}{6}+4\dfrac{2}{5}=$

15. $\dfrac{3}{8} + \dfrac{5}{4} + \dfrac{4}{5} =$

16. $\dfrac{8}{9} - \dfrac{1}{4} =$

17. $\dfrac{7}{10} - \dfrac{3}{14} =$

18. $6\dfrac{1}{3} - 2\dfrac{7}{11} =$

19. $5\dfrac{5}{6} - 3\dfrac{1}{15} =$

20. $\dfrac{2}{13} \times \dfrac{4}{7} =$

21. $\dfrac{5}{6} \times 2\dfrac{2}{5} =$

22. $1\dfrac{1}{8} \times 4\dfrac{3}{4} =$

23. $\dfrac{5}{9} \div \dfrac{10}{13} =$

24. $6\dfrac{2}{3} \div 2\dfrac{1}{6} =$

25. $5\dfrac{3}{10} \div \dfrac{2}{9} =$

Answers

1. Denominator.

   This is the bottom number of any fraction.

2. $\dfrac{2}{7}$

Divide the numerator and denominator by 3.

3. $\dfrac{9}{11}$

Divide the numerator and denominator by 4.

4. $4\dfrac{1}{2}$

9 divided by 2 yields 4 with a remainder of 1.

5. $5\dfrac{3}{5}$

28 divided by 5 yields 5 with a remainder of 3.

6. $\dfrac{10}{3}$

The numerator becomes $3 \times 3 + 1 = 10$.
The denominator remains as 3.

7. $\dfrac{19}{8}$

The numerator becomes $2 \times 8 + 3 = 19$.
The denominator remains as 8.

8. $\dfrac{9}{13}$

Add numerators only.

9. $1\dfrac{1}{10}$

$$\dfrac{3}{5} + \dfrac{1}{2} = \dfrac{6}{10} + \dfrac{5}{10} = \dfrac{11}{10} = 1\dfrac{1}{10}$$

10. $\dfrac{3}{5}$

$$\frac{4}{15}+\frac{1}{3}=\frac{4}{15}+\frac{5}{15}=\frac{9}{15}=\frac{3}{5}$$

11. $>$

$$\frac{2}{3}=\frac{26}{39}\text{ and }\frac{8}{13}=\frac{24}{39}$$

$$26>24$$

12. $=$

$$\frac{12}{16}=\frac{3}{4}\text{ and }\frac{15}{20}=\frac{3}{4}$$

13. $3\dfrac{55}{63}$

$$1\frac{3}{7}=\frac{10}{7},2\frac{4}{9}=\frac{22}{9}$$

The common denominator is 63.

$$\frac{10}{7}+\frac{22}{9}=\frac{90}{63}+\frac{154}{63}=\frac{244}{63}=3\frac{55}{63}$$

14. $8\dfrac{7}{30}$

$$3\frac{5}{6}=\frac{23}{6},4\frac{2}{5}=\frac{22}{5}$$

The common denominator is 30.

$$\frac{23}{6}+\frac{22}{5}=\frac{115}{30}+\frac{132}{30}=\frac{247}{30}=8\frac{7}{30}$$

15. $2\dfrac{17}{40}$

The common denominator is 40.

$$\frac{3}{8}=\frac{15}{40},\frac{5}{4}=\frac{50}{40},\frac{4}{5}=\frac{32}{40}$$

$$\frac{15}{40}+\frac{50}{40}+\frac{32}{40}=\frac{97}{40}=2\frac{17}{40}$$

16. $\dfrac{23}{36}$

The common denominator is 36.

$$\frac{8}{9}=\frac{32}{36},\frac{1}{4}=\frac{9}{36}$$

$$\frac{32}{36}-\frac{9}{36}=\frac{23}{36}$$

17. $\dfrac{17}{35}$

The common denominator is 70.

$$\frac{7}{10}=\frac{49}{70},\frac{3}{14}=\frac{15}{70}$$

$$\frac{49}{70}-\frac{15}{70}=\frac{34}{70},$$

which reduces to $\dfrac{17}{35}$.

18. $3\dfrac{23}{33}$

$$6\frac{1}{3}=\frac{19}{3},2\frac{7}{11}=\frac{29}{11}$$

The common denominator is 33.

$$\frac{19}{3}=\frac{209}{33},\frac{29}{11}=\frac{87}{33}$$

$$\frac{209}{33}-\frac{87}{33}=\frac{122}{33},$$

which reduces to $3\frac{23}{33}$.

19. $2\frac{23}{30}$

$$5\frac{5}{6} = \frac{35}{6}, 3\frac{1}{15} = \frac{46}{15}$$

The common denominator is 30.

$$\frac{35}{6} = \frac{175}{30}, \frac{46}{15} = \frac{92}{30}$$

$$\frac{175}{30} - \frac{92}{30} = \frac{83}{30},$$

which reduces to $2\frac{23}{30}$.

20. $\frac{8}{91}$

Multiply separately the numerators and denominators. No cancellation is possible.

21. 2

$$\frac{5}{6} \times \frac{12}{5},$$

cancel 5's, change 6 to 1, change 12 to 2.

$$\frac{1}{1} \times \frac{2}{1} = 2$$

22. $5\frac{11}{32}$

$$\frac{9}{8} \times \frac{19}{4}$$

Multiply the numerators and denominators separately. No further reduction is possible.

$$\frac{171}{32} = 5\frac{11}{32}$$

23. $\frac{13}{18}$

Change to

$$\frac{5}{9} \times \frac{13}{10}$$

Change 5 to 1, change 10 to 2.

$$\frac{1}{9} \times \frac{13}{2} = \frac{13}{18}$$

24. $3\frac{1}{13}$

Change to

$$\frac{20}{3} \div \frac{13}{6} = \frac{20}{3} \times \frac{6}{13}$$

Change 3 to 1, change 6 to 2.

$$\frac{20}{1} \times \frac{2}{13} = \frac{40}{13} = 3\frac{1}{13}$$

25. $23\frac{17}{20}$

Change to

$$\frac{53}{10} \times \frac{9}{2} = \frac{477}{20} = 23\frac{17}{20}$$

REVIEW

Fractions are another way of viewing parts of whole numbers and decimals. For example, the whole number 7 is 7.0 in decimal form, and written 7/1 as a fraction.

When dealing with fractions, the top part of the fraction is called the numerator and the bottom half is called the denominator. For example, in the fraction 4/7, four is the numerator, and seven is the denominator. Keep in mind that any whole number is really a fraction. The whole number 84, for ins' ce, is really 84/1, which is the same number i ʾaction form.

Fractions may be "simp d" or reduced to the lowest terms by dividin е numerator and denominator of common fa 's. A common factor is any number that can found in both halves of the fraction. For inst e, 2/10, can be simplified to 1/5, since five is mon to both halves of the fraction. (One-t and two-tenths are equal to the same amoun.

Improper fractions are fractions that tain higher numerators than denominato (One example is 59/2.) An improper fractio can be changed to a mixed number, which consists of a whole number and its fractional part. An example of a mixed number is $8^4/_{17}$. A mixed number can also be expressed as an improper fraction; for instance, $1^2/_3 = 5/3$.

Addition or subtraction of fractions with like denominators is done by adding or subtracting the numerators only. The resulting fraction may have to be reduced. This can be seen in this example: 1/12 + 7/12 = 8/12, when reduced, this becomes 2/3. Addition or subtraction of fractions with unlike denominators is done by first finding the lowest common denominator. Each fraction is then converted to a fraction with this denominator. Now you may proceed with the function as you normally would if the denominators were originally the same. The test fraction method can aid in finding the lowest common denominator, also known as the LCD. Keep in mind that any fractions can be compared by size by converting them to fractions with LCD's. Mixed fractions can be added by adding whole parts and fractional parts separately, or by converting them to improper fractions.

Fractions are multiplied by separately multiplying numerators, then denominators. The result should be checked for reducing to lowest terms. Mixed fractions are first converted to improper fractions before multiplying. Cancellation of factors of fractions can be performed before multiplying the fractions. For example, $4/5 \times 1^2/_3 = 4/5 \times 5/3$. The two fives may be canceled so that the result is $4/3 = 1^1/_3$. When dividing fractions, the second fraction is first inverted. The problem is then treated as if it were originally multiplication.

Mathematics

Decimals

MATHEMATICS

DECIMALS

INTRODUCTION TO DECIMALS

As we go through life we learn that not all numbers are whole numbers. In some cases, we have "pieces" of a whole number, for example, when we work with money. When visiting a store, we often find prices like $4.99, $10.35, and $1.79. In each of these cases, we have **decimal numbers**. The digit or digits to the left of the decimal point (.) is the whole number part of our price and the digit or digits to the right of the **decimal point** is the **decimal part**. Working with decimals is very similar to working with whole numbers. Decimal numbers also have place values in the same way as whole numbers. The whole number part of a decimal number has the same place values as we discussed in the section. The place values for the decimal part are very similar except that, instead of the place values ending in "s," the **decimal place** values end in "ths." If we consider the decimal number below, we can see an example of this.

15.3864

1 is in the tens place

5 is in the ones place

3 is in the tenths place

8 is in the hundredths place

6 is in the thousandths place

4 is in the ten thousandths place

The only confusing part of this discussion is that there is no "oneths" or "onths" place. Except for that, the place values are exactly the same. If we pair the decimal point with the ones place, we can see the similarities better—tens and tenths, hundreds and hundredths, etc.

Example 1

Write the place value of the "7" in the following numbers.

(a) 29.7̲52

This "7" is in the <u>tenths</u> place.

(b) 6.127̲5

This "7" is in the <u>thousandths</u> place.

(c) 5.27̲3

This "7" is in the <u>hundredths</u> place.

(d) 0.6597̲

This "7" is in the <u>ten thousandths</u> place.

We can also use this idea of place values to write decimal numbers out in words. In order to do this, we read the decimal point as "and." The number 9.15 would be read as "nine and fifteen hundredths." The number "three hundred fifty-seven thousandths" could be written as 0.357. As we can see, writing decimal numbers in words is very similar to writing whole numbers in words.

Example 2

Write the following numbers in words.

(a) 2.95

Two and ninety-five hundredths

(b) 1.438

One and four hundred thirty-eight thousandths

(c) 7.2

Seven and two tenths

(d) 11.006

Eleven and six thousandths

Example 3

Write the following numbers using digits.

(a) Four and twelve hundredths

4.12

(b) Twelve and seven hundred twenty-six thousandths

12.726

(c) Five and nine tenths

5.9

(d) Six hundred thirty-seven and two thousandths

637.002

As with whole numbers, we can compare decimal numbers. Two numbers can compare in one of the following three ways: =, <, or >. Below are a few examples of comparing decimal numbers.

Example 4

Compare the following pairs of numbers. Write the appropriate symbol between the two numbers.

(a) 5.29 _____ 4.29

We know that 5.29 is a larger number because, moving from left to right, the first digit difference is in the ones position, and 5 is larger than 4.

(b) 0.26 _____ 0.35

We know that 0.35 is a larger number because, moving from left to right, the first digit difference is in the tenths position, and 3 is larger than 2.

(c) 0.1657 _____ 0.1653

We know that 0.1657 is a larger number because, moving from left to right, the first digit difference is in the ten thousandths position, and 7 is larger than 3.

(d) 3.918 _____ 3.908

We know that 3.918 is a larger number because, moving from left to right, the first digit difference is in the hundredths position, and 1 is larger than 0.

Rounding decimal numbers is very similar to rounding whole numbers. Sometimes an approximation can be used instead of working with the exact value. A few examples are listed below.

Example 5

Round the following numbers to the appropriate position.

(a) Round 3.467 to the hundredths position.

3.467—Since the digit to the right is 7, we would add 1 to the marked digit and drop the remainder of the number: 3.47

(b) Round 14.5299 to the tenths position.

14.<u>5</u>299—Since the digit to the right is 2, we would leave the marked digit alone and drop the remainder of the number: 14.5

(c) Round 0.245789 to the ten thousandths position.

0.245<u>7</u>89—Since the digit to the right is 8, we would add 1 to the marked digit and drop the remainder of the number: 0.2458

(d) Round 7.29342 to the thousandths position.

7.29<u>3</u>42—Since the digit to the right is 4, we would leave the marked digit alone and drop the remainder of the number: 7.293

Questions

Write the place value of the underlined digit in the following numbers.

1. 234.67<u>8</u>9

_____

2. 8.<u>1</u>24

_____

3. 0.7640<u>2</u>

_____

4. 69.34<u>6</u>56

_____

5. 49.5<u>3</u>561

_____

Write the following numbers in words.

6. 2.685

_____

7. 4.75

_____

8. 38.7

_____

9. 12.0009

_____

10. 0.456

_____

Write the following numbers using digits.

11. Six and one hundred forty-two thousandths

_____

12. One and five hundredths

_____

13. Ten and six hundred nine thousandths

_____

14. Four and three tenths

_____

15. Twenty-three hundredths

_____

Compare the following pairs of numbers. Write <, =, or > between the two numbers.

16. 5.74 _____ 6.79

17. 2.364 _____ 2.358

18. 0.1235 _____ 0.1235

19. 4.7878 _____ 4.8787

20. 0.00003 _____ 0.000003

Round the following numbers to the underlined position.

21. 56.54_7_8 _____

22. 0.4_8_36 _____

23. 7.5831_3_4 _____

24. 67.528_4_5 _____

25. 0._8_64 _____

Answers

1. Thousandths. The third place value to the right of the decimal point is the thousandths place.

2. Tenths. The first place value to the right of the decimal point is the tenths place.

3. Hundred thousandths. The fifth place value to the right of the decimal point is the hundred thousandths place.

4. Thousandths. The third place value to the right of the decimal point is the thousandths place.

5. Hundredths. The second place value to the

right of the decimal point is the hundredths place.

6. Two and six hundred eighty-five thousandths

7. Four and seventy-five hundredths

8. Thirty-eight and seven tenths

9. Twelve and nine ten thousandths

10. Four hundred fifty-six thousandths

11. 6.142

12. 1.05

13. 10.609

14. 4.3

15. 0.23

16. < 5.74 is less than 6.79.

17. > 2.364 is greater than 2.358.

18. = The two numbers are equal.

19. < 4.7878 is less than 4,8787.

20. > 0.00003 is greater than 0.000003.

21. 56.54_8_

22. 0.4_8_

23. 7.58313_3_

24. 67.528_5_

25. 0._9_

ADDITION OF DECIMALS

Now that we have a better understanding of decimals, what can we do with them? The same four operations (addition, subtraction, multiplication, and division) can be used here. The following sections will deal with each of these operations in turn.

In order to add decimal numbers, we must write the numbers in a column with the decimal points lined up. To help with spacing and columns, we can use zero (0) as a place holder to block off the problem. The examples below demonstrate this method.

Example 1

Add the following pairs of numbers.

(a) Add 2.314 and 4.56.

$$
\begin{array}{r}
2.314 \\
+\ 4.56 \\
\hline
\end{array}
\longrightarrow
\begin{array}{r}
2.314 \\
+\ 4.560 \\
\hline
6.874
\end{array}
$$

(b) Add 23.57 and 3.4848.

$$
\begin{array}{r}
\\
23.57 \\
+\ 3.4848 \\
\hline
\end{array}
\longrightarrow
\begin{array}{r}
11 \\
23.5700 \\
+\ 3.4848 \\
\hline
27.0548
\end{array}
$$

(c) Add 2.217 and 0.489.

$$
\begin{array}{r}
\\
2.217 \\
+\ 0.489 \\
\hline
\end{array}
\longrightarrow
\begin{array}{r}
11 \\
2.217 \\
+\ 0.489 \\
\hline
2.706
\end{array}
$$

(d) Add 3.54327 and 1.14

$$
\begin{array}{r}
3.54327 \\
+\ 1.14 \\
\hline
\end{array}
\longrightarrow
\begin{array}{r}
3.54327 \\
+\ 1.14000 \\
\hline
4.68327
\end{array}
$$

Example 2

Add the following numbers.

(a) Add 2.467482, 1.4126, and 3.46571.

$$
\begin{array}{r}
\\
2.467482 \\
1.4126 \\
+\ 3.46571 \\
\hline
\end{array}
\longrightarrow
\begin{array}{r}
1111 \\
2.467482 \\
1.412600 \\
+\ 3.465710 \\
\hline
7.345792
\end{array}
$$

(b) Add 9.56, 3.34567, and 1.318942.

$$
\begin{array}{r}
\\
9.56 \\
3.34567 \\
+\ 1.318942 \\
\hline
\end{array}
\longrightarrow
\begin{array}{r}
11111 \\
9.560000 \\
3.345670 \\
+\ 1.318942 \\
\hline
14.224612
\end{array}
$$

(c) Add 7.5189, 5.789, 1.7826, and 2.54799.

$$
\begin{array}{r}
\\
7.5189 \\
5.789 \\
1.7826 \\
+\ 2.54799 \\
\hline
\end{array}
\longrightarrow
\begin{array}{r}
2\ \ \ 22 \\
7.51890 \\
5.78900 \\
1.78260 \\
+\ 2.54799 \\
\hline
17.63849
\end{array}
$$

(d) Add 34.867, 43.7891, and 4.33642.

$$
\begin{array}{r}
\\
34.867 \\
43.7891 \\
+\ 4.33642 \\
\hline
\end{array}
\longrightarrow
\begin{array}{r}
1112 \\
34.86700 \\
43.78910 \\
+\ 4.33642 \\
\hline
82.99252
\end{array}
$$

Example 3

Simplify the following problems.

(a)   2.45789 + 3.483

$$
\begin{array}{r}
\underline{11} \\
2.45789 \\
+\ 3.48300 \\
\hline
5.94089
\end{array}
$$

2.45789
+ 3.483 ⟶

(b)   8.45796 + 3.4346 + 43.054

$$
\begin{array}{r}
\underline{1\ 111} \\
8.45796 \\
3.43460 \\
+\ 43.05400 \\
\hline
54.94656
\end{array}
$$

8.45796
3.4346
+ 43.054 ⟶

(c)   34.6 + 65.856 + 43.68743

$$
\begin{array}{r}
\underline{12\ 11} \\
34.60000 \\
65.85600 \\
+\ 43.68743 \\
\hline
144.14343
\end{array}
$$

34.6
65.856
+ 43.68743 ⟶

(d)   25.8967 + 9.357 + 5.24687

$$
\begin{array}{r}
\underline{21\ 221} \\
25.89670 \\
9.35700 \\
+\ 5.24687 \\
\hline
40.50057
\end{array}
$$

25.8967
9.357
+ 5.24687 ⟶

Questions

Add the following pairs of numbers.

1.   Add 1.241 and 3.4634

2.   Add 4.57457 and 3.4678

3.   Add 27.9458 and 89.31762

4.   Add 3.4896, 8.57046, and 3.489861

5.   Add 4.58268, 3.4860547, and 7.8346

6.   Add 24.589434, 56.7257, and 6.76821

7.   Add 14.564, 46.1045, and 31.46789

8.   3.456789 + 8.90417

9.   8.72566 + 4.5745 + 3.463019

10. 24.5869 + 8.24567 + 452.772

Answers

1.   4.7044

$$
\begin{array}{r}
\underline{1} \\
1.241 \\
+\ 3.4634 \\
\hline
4.7044
\end{array}
$$

2.   8.04237

$$
\begin{array}{r}
\underline{1111} \\
4.57457 \\
+\ 3.4678 \\
\hline
8.04237
\end{array}
$$

3.   117.26342

$$
\begin{array}{r}
\underline{11\ 1\ 1} \\
27.9458 \\
+\ 89.31762 \\
\hline
117.26342
\end{array}
$$

4. 15.549921
    1 2 1 1 1
    3.4896
    8.57046
    + 3.489861
    ———————
    15.549921

5. 15.9033347
    1 2 1 1 1
    4.58268
    3.4860547
    + 7.8346
    ———————
    15.9033347

6. 88.083344
    1 2 1 2 1
    24.589434
    56.7257
    + 6.76821
    ———————
    88.083344

7. 92.13639
    1 1 1 1 1
    14.564
    46.1045
    + 31.46789
    ———————
    92.13639

8. 12.360959
    1 1 1
    3.456789
    + 8.90417
    ———————
    12.360959

9. 16.763179
    1 1 1 1
    8.72566
    4.5745
    + 3.463019
    ———————
    16.763179

10. 485.60457
    1 1 1 1 1
    24.5869
    8.24567
    + 452.772
    ———————
    485.60457

SUBTRACTION OF DECIMALS

As we saw in the previous section, working with decimals is very similar to working with whole numbers. Some of the same ideas that we used in addition also apply to subtraction. We need to begin by lining up the decimal points. Using zeros as place holders can also help keep the columns and place values straight. Recall that sometimes we need to borrow in order to subtract. The examples below demonstrate these topics.

Example 1

Subtract the following pairs of numbers.

(a) $8.7981 - 6.562$

    8.7981   8.7981
    − 6.562   →   − 6.5620
    ——————   ——————
                  2.2361

(b) $24.9786 - 3.74$

    24.9786   24.9786
    − 3.74   →   − 3.7400
    ——————   ——————
                  21.2386

(c)   6.54686 – 1.2445

$$
\begin{array}{r}
6.54686 \\
-\ 1.2445 \\
\hline
\end{array}
\longrightarrow
\begin{array}{r}
6.54686 \\
-\ 1.24450 \\
\hline
5.30236
\end{array}
$$

(d)   99.5678 – 73.4

$$
\begin{array}{r}
99.\,5678 \\
-\ 73.4 \\
\hline
\end{array}
\longrightarrow
\begin{array}{r}
99.5678 \\
-\ 73.4000 \\
\hline
26.1678
\end{array}
$$

Example 2

Subtract the following numbers. Borrow when necessary.

(a)   4.25734 – 3.4857

$$
\begin{array}{r}
4.25734 \\
-\ 3.4857 \\
\hline
\end{array}
\longrightarrow
\begin{array}{r}
\textit{1\ 1\ 1} \\
3.15634 \\
-\ 3.48570 \\
\hline
0.77164
\end{array}
$$

(b)   8.982 – 5.6543

$$
\begin{array}{r}
8.982 \\
-\ 5.6543 \\
\hline
\end{array}
\longrightarrow
\begin{array}{r}
\textit{1\ 1} \\
8.9710 \\
-\ 5.6543 \\
\hline
3.3277
\end{array}
$$

(c)   86.8348 – 34.5786

$$
\begin{array}{r}
86.8348 \\
-\ 34.5786 \\
\hline
\end{array}
\longrightarrow
\begin{array}{r}
\textit{1\ 1} \\
86.7238 \\
-\ 34.5786 \\
\hline
52.2552
\end{array}
$$

(d)   34.7802 – 24.89423

$$
\begin{array}{r}
34.7802 \\
-\ 24.89423 \\
\hline
\end{array}
\longrightarrow
\begin{array}{r}
\textit{1\ 11\ 11} \\
23.67910 \\
-\ 24.89423 \\
\hline
9.88597
\end{array}
$$

Example 3

Simplify the following problems.

(a)   20.3189 – 18.687

$$
\begin{array}{r}
20.3189 \\
-\ 18.687 \\
\hline
\end{array}
\longrightarrow
\begin{array}{r}
\textit{11} \\
19.2189 \\
-\ 18.6870 \\
\hline
1.6319
\end{array}
$$

(b)   45.472 – 4.3867

$$
\begin{array}{r}
45.472 \\
-\ 4.3867 \\
\hline
\end{array}
\longrightarrow
\begin{array}{r}
\textit{111} \\
45.3610 \\
-\ 4.3867 \\
\hline
41.0853
\end{array}
$$

(c)   4.567892 – 1.346789

$$
\begin{array}{r}
4.567892 \\
-\ 1.346789 \\
\hline
\end{array}
\longrightarrow
\begin{array}{r}
\textit{1} \\
4.567882 \\
-\ 1.346789 \\
\hline
3.221103
\end{array}
$$

(d)   13.45823 – 8.907349

$$
\begin{array}{r}
13.45823 \\
-\ 8.907349 \\
\hline
\end{array}
\longrightarrow
\begin{array}{r}
\textit{1\quad 111} \\
12.457120 \\
-\ 8.907349 \\
\hline
4.550881
\end{array}
$$

Questions

Subtract the following numbers.

1. 5.68975 − 2.0346

2. 8.9785 − 7.62

3. 78.9035 − 34.6021

4. 134.61348 − 34.501

5. 78.96759 − 24.524

6. 9.5867 − 4.1571

7. 5.63978 − 1.542

8. 65.876 − 48.1323

9. 97.6538 − 1.23456

10. 45.892459 − 2.35634

11. 41.25896 − 10.2465

12. 12.3415 − 2.63756

13. 98.07455 − 31.613

14. 24.5678 − 16.78679

15. 458.9756 − 245.86789

Answers

1. 3.65515

    ```
      5.68975
    − 2.0346
    ─────────
      3.65515
    ```

2. 1.3585

    ```
      8.9785
    − 7.62
    ────────
      1.3585
    ```

3. 44.3014

    ```
      78.9035
    − 34.6021
    ──────────
      44.3014
    ```

4. 100.11248

    ```
      134.61348
    − 34.501
    ────────────
      100.11248
    ```

5. 54.44359

    ```
      78.96759
    − 24.524
    ───────────
      54.44359
    ```

6. 5.4296

    ```
          1
      9.5867
    − 4.1571
    ─────────
      5.4296
    ```

7. 4.09778

    ```
          1
      5.63978
    − 1.542
    ──────────
      4.09778
    ```

8. 17.7437

    ```
        1 1
      65.8760
    − 48.1323
    ──────────
      17.7437
    ```

9. 96.41924

    1 1

   97.65380

  – 1.23456

   96.41924

10. 43.536119

    1

  45.892459

 – 2.356340

  43.536119

11. 31.01246

   41.25896

 – 10.2465

   31.01246

12. 9.70394

   1 1 111

  12.34150

 – 2.63756

   9.70394

13. 66.46155

    1

  98.07455

 – 31.613

   66.46155

14. 7.78101

   1 11 1

  24.56780

 – 16.78679

   7.78101

15. 213.10771

    111

  458.97560

 – 245.86789

  213.10771

MULTIPLICATION OF DECIMALS

Now that we are "experts" in addition and subtraction of decimals, we can switch our discussion to multiplication. Multiplying decimals is closer to working with whole numbers than either of the above topics. Here we want to write the numbers in a table format and multiply them together as if the decimal points were not present. After we get a result we must take into account that the original problem was made up of decimal numbers. Next, we must add the decimal places in all the numbers involved in the problem. In order to get the proper placement of the decimal point in the final answer, we count off the correct number of places moving right to left. The examples that follow consider most of the details that might arise.

Example 1

Multiply the following pairs of numbers.

(a) 5.68×2.3

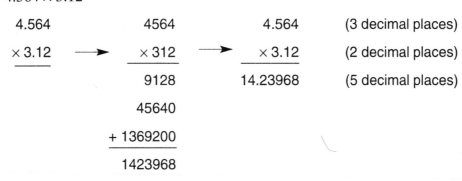

5.68	568	5.68	(2 decimal places)
× 2.3	× 23	× 2.3	(1 decimal place)
	1704	13.064	(3 decimal places)
	+ 11360		
	13064		

(b) 4.564×3.12

4.564	4564	4.564	(3 decimal places)
× 3.12	× 312	× 3.12	(2 decimal places)
	9128	14.23968	(5 decimal places)
	45640		
	+ 1369200		
	1423968		

(c) 34.6178×2.3

34.6178	346178	34.6178	(4 decimal places)
× 2.3	× 23	× 2.3	(1 decimal place)
	1038534	79.62094	(5 decimal places)
	+ 6923560		
	7962094		

(d) 13.466×3.467

13.466	13466	13.466	(3 decimal places)
× 3.467	× 3467	× 3.467	(3 decimal places)
	94262	46.686622	(6 decimal places)
	807960		
	5386400		
	+ 40398000		
	46686622		

Example 2

Multiply the following numbers.

(a) 0.258×0.6

0.258	258	0.258 (3 decimal places)
× 0.6 →	× 6 →	× 0.6 (1 decimal place)
	1548	0.1548 (4 decimal places)

(b) 0.023×0.07

0.023	23	0.023 (3 decimal places)
× 0.07 →	× 7 →	× 0.07 (2 decimal places)
	161	0.00161 (5 decimal places)

(c) 1.7465×0.13

1.7465	17465	1.7465 (4 decimal places)
× 0.13 →	× 13 →	× 0.13 (2 decimal places)
	52395	0.227045 (6 decimal places)
	+ 174650	
	227045	

(d) 7.824×0.15

7.824	7824	7.824 (3 decimal places)
× 0.15 →	× 15 →	× 0.15 (2 decimal places)
	39120	1.17360 (5 decimal places)
	+ 78240	
	117360	

Example 3

Simplify the following problems.

(a) 341.786×6.7

341.786	341786	341.786	(3 decimal places)
$\times 6.7$	$\times 67$	$\times 6.7$	(1 decimal place)
	2392502	2289.9662	(4 decimal places)
	+ 20507160		
	22899662		

(b) 0.6045×3.7

0.6045	6045	0.6045	(4 decimal places)
$\times 3.7$	$\times 37$	$\times 3.7$	(1 decimal place)
	42315	2.23665	(5 decimal places)
	+ 181350		
	223665		

(c) 4579.2×9.01

4579.2	45792	4579.2	(1 decimal place)
$\times 9.01$	$\times 901$	$\times 9.01$	(2 decimal places)
	45792	41258.592	(3 decimal places)
	000000		
	+ 41212800		
	41258592		

(d) 2.545×5.67

2.545	2545	2.545	(3 decimal places)
$\times 5.67$	$\times 567$	$\times 5.67$	(2 decimal places)
	17815	14.43015	(5 decimal places)
	152700		
	+ 1272500		
	1443015		

Questions

Multiply the following numbers.

1. 5.69×25.7

2. 3.568×4.2

3. 78.752×1.36

4. 312.5×0.567

5. 0.9802×34.6

6. 9.673×34.16

7. 0.9674×0.56

8. 4.74×0.75

9. 3.1358×0.512

10. 0.24789×0.13

11. 34.7×0.68

12. 790.4×4.6

13. 2.356×0.971

14. 3.689×7.95

15. 134.689×2.34

Answers

1. 146.233

5.69	(2 decimal places)
× 25.7	(1 decimal place)
3983	
2845	
+ 1138	
146.233	(3 decimal places)

2. 14.9856

3.568	(3 decimal places)
× 4.2	(1 decimal place)
7136	
+ 14272	
14.9856	(4 decimal places)

3. 107.10272

78.752	(3 decimal places)
× 1.36	(2 decimal places)
472512	
236256	
+ 78752	
107.10272	(5 decimal places)

4. 177.1875

312.5	(1 decimal place)
× 0.567	(3 decimal places)
21875	
18750	
+ 15625	
177.1875	(4 decimal places)

5. 33.91492
      0.9802   (4 decimal places)
    × 34.6   (1 decimal place)
     58812
     39208
  + 29406
   33.91492   (5 decimal places)

6. 330.42968
      9.673   (3 decimal places)
   × 34.16   (2 decimal places)
    58038
     9673
    38692
  + 29019
  330.42968   (5 decimal places)

7. 0.541744
    0.9674   (4 decimal places)
   × 0.56   (2 decimal places)
   58044
  + 48370
  0.541744   (6 decimal places)

8. 3.555
    4.74   (2 decimal places)
  × 0.75   (2 decimal places)
   2370
  + 3318
  3.5550   (4 decimal places)

9. 1.6055296
    3.1358   (4 decimal places)
  × 0.512   (3 decimal places)
   62716
   31358
 + 156790
 1.6055296   (7 decimal places)

10. 0.0322257
   0.24789   (5 decimal places)
   × 0.13   (2 decimal places)
   74367
 + 24789
 0.0322257   (7 decimal places)

11. 23.596
    34.7   (1 decimal place)
  × 0.68   (2 decimal places)
   2776
 + 2082
 23.596   (3 decimal places)

12. 3,635.84
   790.4   (1 decimal place)
   × 4.6   (1 decimal place)
  47424
 + 31616
 3635.84   (2 decimal places)

13. 2.287676
   2.356   (3 decimal places)
  × 0.971   (3 decimal places)
   2356
  16492
 + 21204
 2.287676   (6 decimal places)

14. 29.32755

3.689	(3 decimal places)
× 7.95	(2 decimal places)
18445	
33201	
+ 25823	
29.32755	(5 decimal places)

15. 315.17226

134.689	(3 decimal places)
× 2.34	(2 decimal places)
538756	
404067	
+ 269378	
315.17226	(5 decimal places)

DIVISION OF DECIMALS

We divide decimals in essentially the same way that we divided whole numbers. If we are dividing by a whole number, we just bring the decimal point up to our answer line and then divide as we did in the previous chapter. On the other hand, if we are dividing by a decimal number, we must first move the decimal point of this number to the right as many spaces as needed to make it a whole number. In order to keep our problem equivalent, we must move the decimal point of the number under the division bar by the same number of places. If we move the decimal point two spaces to the right in the number outside, we must also move the decimal point two spaces to the right inside the division bar. In the following examples, we will address both of these situations. Instead of leaving our answer with a remainder, we can use zeros as place holders and continue dividing until our answer works out exactly.

Example 1

Divide the following numbers.

(a) $6.412 \div 2$

$$
\begin{array}{r}
. \\
2\overline{)6.412}
\end{array}
\longrightarrow
\begin{array}{r}
3. \\
2\overline{)6.412} \\
-6 \\
\hline
4
\end{array}
\longrightarrow
\begin{array}{r}
3.2 \\
2\overline{)6.412} \\
-6 \\
\hline
4 \\
-4 \\
\hline
1
\end{array}
\longrightarrow
\begin{array}{r}
3.206 \\
2\overline{)6.412} \\
-6 \\
\hline
4 \\
-4 \\
\hline
12 \\
-12 \\
\hline
0
\end{array}
$$

Since we are dividing by a whole number, we just need to move the decimal point up to our answer line and divide as we did in the last section. How many times will 2 go into 6? Our answer here would be 3. Since $2 \times 3 = 6$, we can place this value under the first number and subtract. If we subtract and bring down the next digit, we get 4. Now we need to consider how many times 2 goes into 4. Our answer is 2. Since $2 \times 2 = 4$, we can place this value under the first number and subtract. If we subtract and bring down the next digit, we get 1. Since 2 does not go into 1, we place a zero on our answer line in this place and bring down another digit. Now, consider how many times 2 goes into 12. Our answer is 6. Since $2 \times 6 = 12$, we can place this value under the first number and subtract. Here we get $12 - 12 = 0$. We have no remainder in this problem. Our overall answer would be 3.206.

(b) $5.8956 \div 17$

$$
\begin{array}{r}
. \\
17 \overline{)5.8956}
\end{array}
\longrightarrow
\begin{array}{r}
0.3 \\
17 \overline{)5.8956} \\
-51 \\
\hline
79
\end{array}
\longrightarrow
\begin{array}{r}
0.34 \\
17 \overline{)5.8956} \\
-51 \\
\hline
79 \\
-68 \\
\hline
115
\end{array}
\longrightarrow
$$

$$
\longrightarrow
\begin{array}{r}
0.346 \\
17 \overline{)5.8956} \\
-51 \\
\hline
79 \\
-68 \\
\hline
115 \\
-102 \\
\hline
136
\end{array}
\longrightarrow
\begin{array}{r}
0.3468 \\
17 \overline{)5.8956} \\
-51 \\
\hline
79 \\
-68 \\
\hline
115 \\
-102 \\
\hline
136 \\
-136 \\
\hline
0
\end{array}
$$

Since we are dividing by a whole number, we just need to move the decimal point up to our answer line and divide as we did in the last section. How many times will 17 go into 58? Our answer here would be 3. Since $17 \times 3 = 51$, we can place this value under the first number and subtract. If we subtract and bring down the next digit, we get 79. Now we need to consider how many times 17 goes into 79. Our answer is 4. Since $17 \times 4 = 68$, we can place this value under the first number and subtract. If we subtract and bring down the next digit, we get 115. Now we consider how many times 17 goes into 115. Our answer is 6. Since $17 \times 6 = 102$, we can place this value under the first number and subtract. If we subtract and bring down the next digit, we will get 136. Now we consider how many times 17 goes into 136. Our answer is 8. Since $17 \times 8 = 136$, we can place this value under the first number and subtract. Here we get $136 - 136 = 0$. We have no remainder in this problem. Our overall answer would be 0.3468.

(c) $1218.14 \div 49$

$$
\begin{array}{r}
. \\
49 \overline{\smash)1218.14}
\end{array}
\longrightarrow
\begin{array}{r}
2\ . \\
49 \overline{\smash)1218.14} \\
-98 \\ \hline
238
\end{array}
\longrightarrow
\begin{array}{r}
24. \\
49 \overline{\smash)1218.14} \\
-98 \\ \hline
238 \\
-196 \\ \hline
42\ 1
\end{array}
\longrightarrow
$$

$$
\longrightarrow
\begin{array}{r}
24.8 \\
49 \overline{\smash)1218.14} \\
-98 \\ \hline
238 \\
-196 \\ \hline
42\ 1 \\
-39\ 2 \\ \hline
2\ 94
\end{array}
\longrightarrow
\begin{array}{r}
24.86 \\
49 \overline{\smash)1218.14} \\
-98 \\ \hline
238 \\
-196 \\ \hline
42\ 1 \\
-39\ 2 \\ \hline
2\ 94 \\
-2\ 94 \\ \hline
0
\end{array}
$$

Since we are dividing by a whole number, we just need to move the decimal point up to our answer line and divide as we did in the last section. How many times will 49 go into 121? Our answer is 2. Since $49 \times 2 = 98$, we can place this value under the first number and subtract. If we subtract and bring down the next digit, we will get 238. Now we need to consider how many times 49 goes into 238. Our answer is 4. Since $49 \times 4 = 196$, we can place this value under the first number and subtract. If we subtract and bring down the next digit, we get 421. Now we consider how many times 49 goes into 421. Our answer is 8. Since $49 \times 8 = 392$, we can place this value under the first number and subtract. If we subtract and bring down the next digit, we will get 294. Now we consider how many times 49 goes into 294. Our answer is 6. Since $49 \times 6 = 294$, we can place this value under the first number and subtract. Here we get $294 - 294 = 0$. We have no remainder in this problem. Our overall answer would be 24.86.

(d) $11.82636 \div 156$

```
         .                    0.07                  0.075
156 | 11.82636  →    156 | 11.82636   →   156 | 11.82636  →
                         - 1092                     - 1092
                            906                        906
                                                     - 780
                                                       1263
```

```
                    0.0758                  0.07581
         → 156 | 11.82636   →   156 | 11.82636
                - 1092                     - 1092
                   906                        906
                  - 780                      - 780
                   1263                       1263
                 - 1248                     - 1248
                    156                        156
                                            - 156
                                               0
```

Since we are dividing by a whole number, we just need to move the decimal point up to our answer line and divide as we did in the last section. How many times will 156 go into 1182? Our answer is 7. Since $156 \times 7 = 1092$, we can place this value under the first number and subtract. If we subtract and bring down the next digit, we get 906. Now we need to consider how many times 156 goes into 906. Our answer is 5. Since $156 \times 5 = 780$, we can place this value under the first number and subtract. If we subtract and bring down the next digit, we get 1263. Now we consider how many times 156 goes into 1263. Our answer is 8. Since $156 \times 8 = 1248$, we can place this value under the first number and subtract. If we subtract and bring down the next digit, we will get 156. Now we consider how many times 156 goes into 156. Our answer is 1. Since $156 \times 1 = 156$, we can place this value under the first number and subtract. Here we get $156 - 156 = 0$. We have no remainder in this problem. Our overall answer would be 0.07581.

Example 2

Divide the following numbers.

(a) $74.36 \div 0.2$

$$0.2\overline{)74.36}\ \cdot \longrightarrow \quad 2\overline{)743.6}\ \overset{3\ \cdot}{} \longrightarrow \quad 2\overline{)743.6}\ \overset{37\ \cdot}{} \longrightarrow$$

First step:
$$\begin{array}{r} 3\ \cdot \\ 2\overline{)743.6} \\ -6 \\ \hline 14 \end{array}$$

Second step:
$$\begin{array}{r} 37\ \cdot \\ 2\overline{)743.6} \\ -6 \\ \hline 14 \\ -14 \\ \hline 3 \end{array}$$

$$\longrightarrow \quad \begin{array}{r} 371. \\ 2\overline{)743.6} \\ -6 \\ \hline 14 \\ -14 \\ \hline 3 \\ -2 \\ \hline 1\,6 \end{array} \quad \longrightarrow \quad \begin{array}{r} 371.8 \\ 2\overline{)743.6} \\ -6 \\ \hline 14 \\ -14 \\ \hline 3 \\ -2 \\ \hline 1\,6 \\ -1\,6 \\ \hline 0 \end{array}$$

We are familiar with dividing a number by 3 or 7, but not by 0.2. By moving the decimal point one place to the right, this value becomes 2. In order to keep this problem equivalent, we must also move the decimal inside the division bar one space to the right. This changes 74.36 into 743.6. Now we divide 2 into 743.6 as we did in the examples above. Two goes into 7 three times. Since $2 \times 3 = 6$, we can place this value under the first number and subtract. This gives 1, and by bringing down the next digit, we get 14. Two goes into 14 seven times. Since $2 \times 7 = 14$, we can place this value under the first number and subtract. This gives 0. When we bring down the next digit, we get 3. Two goes into 3 once. Since $2 \times 1 = 2$, we can place this value under the first number and subtract. This gives 1. When we bring down the next digit, we get 16. Two goes into 16 eight times. Since $2 \times 8 = 16$, we can place this value under the first number and subtract. This gives us 0. Our overall answer is 371.8.

(b) $0.876 \div 0.06$

$$
0.06\overline{)0.876} \quad\longrightarrow\quad
\begin{array}{r}
1\ . \\
6\,\overline{)87.6} \\
-6 \\ \hline
27
\end{array}
\quad\longrightarrow\quad
\begin{array}{r}
14. \\
6\,\overline{)87.6} \\
-6 \\ \hline
27 \\
-24 \\ \hline
36
\end{array}
\quad\longrightarrow\quad
\begin{array}{r}
14.6 \\
6\,\overline{)87.6} \\
-6 \\ \hline
27 \\
-24 \\ \hline
36 \\
-36 \\ \hline
0
\end{array}
$$

Since we are dividing by 0.06, we must move the decimal point two places to the right. In order to keep this problem equivalent, we must also move the decimal point under the division bar two places to the right. This changes 0.876 into 87.6. Six goes into 8 once. Since $6 \times 1 = 6$, we can place this value under the first number and subtract. This gives 2. When we bring down the next digit, we get 27. Six goes into 27 four times. Since $6 \times 4 = 24$, we can place this value under the first number and subtract. This will give us 3. When we bring down the next digit, we get 36. Six goes into 36 six times. Since $6 \times 6 = 36$, we can place this value under the first number and subtract. This gives us 0. Therefore, our final answer is 14.6.

(c) $6.1663 \div 0.23$

$$
0.23\overline{)6.1663} \quad\longrightarrow\quad
\begin{array}{r}
2\ . \\
23\,\overline{)616.63} \\
-46 \\ \hline
156
\end{array}
\quad\longrightarrow\quad
\begin{array}{r}
26. \\
23\,\overline{)616.63} \\
-46 \\ \hline
156 \\
-138 \\ \hline
18\,6
\end{array}
\quad\longrightarrow
$$

$$
\longrightarrow\quad
\begin{array}{r}
26.8 \\
23\,\overline{)616.63} \\
-46 \\ \hline
156 \\
-138 \\ \hline
18\,6 \\
-18\,4 \\ \hline
23
\end{array}
\quad\longrightarrow\quad
\begin{array}{r}
26.81 \\
23\,\overline{)616.63} \\
-46 \\ \hline
156 \\
-138 \\ \hline
18\,6 \\
-18\,4 \\ \hline
23 \\
-23 \\ \hline
0
\end{array}
$$

Since we are dividing by 0.23, we must move the decimal point two places to the right. In order to keep this problem equivalent, we must also move the decimal point under the division bar two places to the right. This changes 6.1663 into 616.63. Twenty-three goes into 61 twice. Since $23 \times 2 = 46$, we can place this value under the first number and subtract. This gives 15, and when we bring down the next digit, we get 156. Twenty-three goes into 156 six times. Since $23 \times 6 = 138$, we can place this value under the first number and subtract. This will give us 18, and when we bring down the next digit, we get 186. Twenty-three goes into 186 eight times. Since $23 \times 8 = 184$, we can place this value under the first number and subtract. This gives us 2, and when we bring down the next digit, we get 23. Twenty-three goes into 23 once. Since $23 \times 1 = 23$, we can place this value under the first number and subtract. This gives us 0. Therefore, our final answer is 26.81.

(d) $22.5504 \div 5.4$

$$
\begin{array}{r}
5.4\,\overline{)22.5504}
\end{array}
\longrightarrow
\begin{array}{r}
4. \\
54\,\overline{)225.504} \\
-216 \\
\hline
9\,5
\end{array}
\longrightarrow
\begin{array}{r}
4.1 \\
54\,\overline{)225.504} \\
-216 \\
\hline
9\,5 \\
-5\,4 \\
\hline
4\ 10
\end{array}
\longrightarrow
$$

$$
\longrightarrow
\begin{array}{r}
4.17 \\
54\,\overline{)225.504} \\
-216 \\
\hline
9\,5 \\
-5\,4 \\
\hline
4\ 10 \\
-3\,78 \\
\hline
324
\end{array}
\longrightarrow
\begin{array}{r}
4.176 \\
54\,\overline{)225.504} \\
-216 \\
\hline
9\,5 \\
-5\,4 \\
\hline
4\ 10 \\
-3\,78 \\
\hline
324 \\
-324 \\
\hline
0
\end{array}
$$

Since we are dividing by 5.4, we must move the decimal point one place to the right. In order to keep this problem equivalent, we must also move the decimal point under the division bar one place to the right. This changes 22.5504 into 225.504. Fifty-four goes into 225 four times. Since $54 \times 4 = 216$, we can place this value under the first number and subtract. This gives 9, and when we bring down the next digit, we get 95. Fifty-four goes into 95 once. Since $54 \times 1 = 54$, we can place this value under the first number and subtract. This will give us 41, and when we bring down the next digit, we get 410. Fifty-four goes into 410 seven times. Since $54 \times 7 = 378$, we can place this value under the first number and subtract. This gives us 32, and when we bring down the next digit, we get 324. Fifty-four goes into 324 six times. Since $54 \times 6 = 324$, we can place this value under the first number and subtract. This gives us 0. Therefore, our overall answer is 4.176.

Example 3

Divide the following numbers.

(a) $7.319 \div 5$

$$
\begin{array}{r}
. \\
5\,\overline{)7.319}
\end{array}
\longrightarrow
\begin{array}{r}
1. \\
5\,\overline{)7.319} \\
-5 \\
\hline
2\,3
\end{array}
\longrightarrow
\begin{array}{r}
1.4 \\
5\,\overline{)7.319} \\
-5 \\
\hline
2\,3 \\
-2\,0 \\
\hline
31
\end{array}
\longrightarrow
\begin{array}{r}
1.46 \\
5\,\overline{)7.319} \\
-5 \\
\hline
2\,3 \\
-2\,0 \\
\hline
31 \\
-30 \\
\hline
19
\end{array}
\longrightarrow
$$

$$
\longrightarrow
\begin{array}{r}
1.463 \\
5\,\overline{)7.319} \\
-5 \\
\hline
2\,3 \\
-2\,0 \\
\hline
31 \\
-30 \\
\hline
19 \\
-15 \\
\hline
4
\end{array}
\longrightarrow
\begin{array}{r}
1.4638 \\
5\,\overline{)7.3190} \\
-5 \\
\hline
2\,3 \\
-2\,0 \\
\hline
31 \\
-30 \\
\hline
19 \\
-15 \\
\hline
40 \\
-40 \\
\hline
0
\end{array}
$$

In this example, we are dividing 5 into 7.319. Since we are dividing by a whole number, we do not have to move any of the decimal points. The decimal point of the number under the division bar comes straight up to our answer line. Here, 5 goes into 7 once. Since $5 \times 1 = 5$, we can place this value under the first number and subtract. This gives 2, and when we bring down the next digit, we get 23. Five goes into 23 four times. Since $5 \times 4 = 20$, we can place this value under the first number and subtract. This will give 3, and when we bring down the next digit, we get 31. Five goes into 31 six times. Since $5 \times 6 = 30$, we can place this value under the first number and subtract. This gives us 1, and when we bring down the next digit, we get 19. Five goes into 19 three times. Since $5 \times 3 = 15$, we can place this value under the first number and subtract. This will give 4. Since we do not want to write our answer with a remainder, we can write a zero as a place holder at the end of the number and continue the division. This is the italicized zero (*0*). Five goes into 40 eight times. Since $5 \times 8 = 40$, we can place this value under the first number and subtract. This gives us 0. Our overall answer is 1.4638.

(b) $1.014 \div 1.2$

$$1.2\,\overline{)\,1.014} \;\longrightarrow\; 12\,\overline{)\,10.14} \;\longrightarrow\; 12\,\overline{)\,10.14} \;\longrightarrow\; 12\,\overline{)\,10.140}$$

	0.8	0.84	0.845
	-96	-96	-96
	54	54	54
		-48	-48
		6	60
			-60
			0

In order to divide 1.2 into 1.014, we must move the decimal points one place to the right. This gives us 12 being divided into 10.14. Twelve goes into 101 eight times. Since $12 \times 8 = 96$, we can place this value under the first number and subtract. This gives 5, and when we bring down the next digit, we get 54. Twelve goes into 54 four times. Since $12 \times 4 = 48$, we can place this value under the first number and subtract. This gives 6. Again, in this example, we can write a zero as a place holder at the end of the number and continue our division. This is the italicized zero (*0*) in the problem. Twelve goes into 60 five times. Since $12 \times 5 = 60$, we can place this value under the first number and subtract. This gives 0. Our final answer is 0.845.

(c) $0.2235 \div 0.06$

$$0.06\,\overline{)\,0.2235} \;\longrightarrow\; 6\,\overline{)\,22.35} \;\longrightarrow\; 6\,\overline{)\,22.35} \;\longrightarrow$$

	3.	3.7
	-18	-18
	43	43
		-42
		15

$$\longrightarrow\; 6\,\overline{)\,22.35} \;\longrightarrow\; 6\,\overline{)\,22.350}$$

	3.72	3.725
	-18	-18
	43	43
	-42	-42
	15	15
	-12	-12
	3	30
		-30
		0

Here, we are dividing 0.2235 by 0.06. In order to do this, we must move our decimal points two places to the right. We now have 6 being divided into 22.35. Six goes into 22 three times. Since $6 \times 3 = 18$, we can place this value under the first number and subtract. This will give us 4, and when we bring down the next digit, we get 43. Six goes into 43 seven times. Since $6 \times 7 = 42$, we can place this value under the first number and subtract. This gives 1, and when we bring down the next digit, we get 15. Six goes into 15 twice. Since $6 \times 2 = 12$, we can place this value under the first number and subtract. This will give us 3. If we place a zero after the number and continue our division, we can work this problem out exactly. Six goes into 30 five times. Since $6 \times 5 = 30$, we can place this value under the first number and subtract. This gives us 0. Therefore, our final answer is 3.725.

(d) $1.8629 \div 0.25$

$$
\begin{array}{r}
. \\
0.25\,\overline{)1.8629}
\end{array}
\longrightarrow
\begin{array}{r}
7. \\
25\,\overline{)186.29} \\
-175 \\
\hline
11\,2
\end{array}
\longrightarrow
\begin{array}{r}
7.4 \\
25\,\overline{)186.29} \\
-175 \\
\hline
11\,2 \\
-10\,0 \\
\hline
1\,29
\end{array}
\longrightarrow
$$

$$
\longrightarrow
\begin{array}{r}
7.45 \\
25\,\overline{)186.29} \\
-175 \\
\hline
11\,2 \\
-10\,0 \\
\hline
1\,29 \\
-1\,25 \\
\hline
4
\end{array}
\longrightarrow
\begin{array}{r}
7.451 \\
25\,\overline{)186.29\mathit{0}} \\
-175 \\
\hline
11\,2 \\
-10\,0 \\
\hline
1\,29 \\
-1\,25 \\
\hline
40 \\
-25 \\
\hline
15
\end{array}
\longrightarrow
\begin{array}{r}
7.4516 \\
25\,\overline{)186.29\mathit{00}} \\
-175 \\
\hline
11\,2 \\
-10\,0 \\
\hline
1\,29 \\
-1\,25 \\
\hline
40 \\
-25 \\
\hline
15\mathit{0} \\
-150 \\
\hline
0
\end{array}
$$

In order to divide 1.8629 by 0.25, we must move the decimal points two places to the right. Now we are dividing 186.29 by 25. Twenty-five goes into 186 seven times. Since $25 \times 7 = 175$, we can place this value under the first number and subtract. This gives 11, and when we bring down the next digit, we get 112. Twenty-five goes into 112 four times. Since $25 \times 4 = 100$, we can place this value under the first number and subtract. This gives 12, and when we bring down the next digit, we get 129. Twenty-five goes into 129 five times. Since $25 \times 5 = 125$, we can place this value under the first number and subtract. This gives us 4, and if we bring down the next digit (a zero used as a place

holder at the end of the number), we get 40. Twenty-five goes into 40 once. Since 25 ×
1 = 25, we can place this value under the first number and subtract. This gives us 15,
and when we bring down the next digit (another zero used as a place holder at the end
of the number), we get 150. Twenty-five goes into 150 six times. Since 25 × 6 = 150,
we can place this value under the first number and subtract. This gives us 0. Our final
answer is 7.4516.

Questions

Divide the following numbers.

1. $9.6312 \div 3$

2. $1{,}135.52 \div 47$

3. $52{,}151.4 \div 126$

4. $59.276 \div 0.7$

5. $4.3505 \div 3.5$

6. $35.075 \div 0.61$

7. $7.06 \div 5$

8. $163.1463 \div 4.3$

9. $0.348 \div 0.16$

10. $0.941196 / 0.123$

Answers

1. 3.2104

```
      3.            3.2            3.21           3.210          3.2104
  3 │9.6312    3 │9.6312      3 │9.6312      3 │9.6312      3 │9.6312
   -9            -9             -9             -9             -9
   ─────         ─────          ─────          ─────          ─────
   0 6           0 6            0 6            0 6            0 6
                 -6             -6             -6             -6
                 ─────          ─────          ─────          ─────
                 03             03             03             03
                                -3             -3             -3
                                ─────          ─────          ─────
                                01             01             01
                                               -0             -0
                                               ─────          ─────
                                               12             12
                                                              -12
                                                              ─────
                                                               0
```

2. 24.16

```
       2 .              24 .              24.1              24.16
 47 │1135.52      47 │1135.52       47 │1135.52       47 │1135.52
    -94              -94               -94               -94
    ─────            ─────             ─────             ─────
    195              195               195               195
                     -188              -188              -188
                     ─────             ─────             ─────
                     75                75                75
                                       -47               -47
                                       ─────             ─────
                                       282               282
                                                         -282
                                                         ─────
                                                          0
```

3. 413.9

```
        4 .                41 .                413.               413.9
 126 │52151.4      126 │52151.4       126 │52151.4       126 │52151.4
     -504              -504               -504               -504
     ─────            ─────             ─────             ─────
     175              175               175               175
                      -126              -126              -126
                      ─────             ─────             ─────
                      49                491               491
                                        -378              -378
                                        ─────             ─────
                                        1134              1134
                                                          -1134
                                                          ─────
                                                           0
```

4. 84.68

$$0.7\overline{)59.276} \longrightarrow 7\overline{)\begin{matrix} 8. \\ 592.76 \end{matrix}} \longrightarrow 7\overline{)\begin{matrix} 84. \\ 592.76 \end{matrix}} \longrightarrow 7\overline{)\begin{matrix} 84.6 \\ 592.76 \end{matrix}} \longrightarrow 7\overline{)\begin{matrix} 84.68 \\ 592.76 \end{matrix}}$$

		8.		84.		84.6		84.68
		− 56		− 56		− 56		− 56
		32		32		32		32
				− 28		− 28		− 28
				47		47		47
						− 42		− 42
						56		56
								− 56
								0

5. 1.243

$$3.5\overline{)4.3505} \longrightarrow 35\overline{)\begin{matrix} 1. \\ 43.505 \end{matrix}} \longrightarrow 35\overline{)\begin{matrix} 1.2 \\ 43.505 \end{matrix}} \longrightarrow 35\overline{)\begin{matrix} 1.24 \\ 43.505 \end{matrix}} \longrightarrow 35\overline{)\begin{matrix} 1.243 \\ 43.505 \end{matrix}}$$

	1.		1.2		1.24		1.243
	− 35		− 35		− 35		− 35
	85		85		85		85
			− 70		− 70		− 70
			150		150		150
					− 140		− 140
					105		105
							− 105
							0

6. 57.5

$$0.61\overline{)35.075} \longrightarrow 0.61\overline{)\begin{matrix} 5. \\ 3507.5 \end{matrix}} \longrightarrow 0.61\overline{)\begin{matrix} 57. \\ 3507.5 \end{matrix}} \longrightarrow 0.61\overline{)\begin{matrix} 57.5 \\ 3507.5 \end{matrix}}$$

	5.		57.		57.5
	− 305		− 305		− 305
	457		457		457
			− 427		− 427
			305		305
					− 305
					0

7. 1.412

$$
\begin{array}{r} 1. \\ 5\,\overline{)\,7.06} \\ -5 \\ \hline 20 \end{array}
\longrightarrow
\begin{array}{r} 1.4 \\ 5\,\overline{)\,7.06} \\ -5 \\ \hline 20 \\ -20 \\ \hline 06 \end{array}
\longrightarrow
\begin{array}{r} 1.41 \\ 5\,\overline{)\,7.060} \\ -5 \\ \hline 20 \\ -20 \\ \hline 06 \\ -5 \\ \hline 10 \end{array}
\longrightarrow
\begin{array}{r} 1.412 \\ 5\,\overline{)\,7.060} \\ -5 \\ \hline 20 \\ -20 \\ \hline 06 \\ -5 \\ \hline 10 \\ -10 \\ \hline 0 \end{array}
$$

8. 37.941

$$
\begin{array}{r} . \\ 4.3\,\overline{)\,163.1463} \end{array}
\longrightarrow
\begin{array}{r} 3\ . \\ 43\,\overline{)\,1631.463} \\ -129 \\ \hline 341 \end{array}
\longrightarrow
\begin{array}{r} 37. \\ 43\,\overline{)\,1631.463} \\ -129 \\ \hline 341 \\ -301 \\ \hline 404 \end{array}
\longrightarrow
$$

$$
\longrightarrow
\begin{array}{r} 37.9 \\ 43\,\overline{)\,1631.463} \\ -129 \\ \hline 341 \\ -301 \\ \hline 404 \\ -387 \\ \hline 176 \end{array}
\longrightarrow
\begin{array}{r} 37.94 \\ 43\,\overline{)\,1631.463} \\ -129 \\ \hline 341 \\ -301 \\ \hline 404 \\ -387 \\ \hline 176 \\ -172 \\ \hline 43 \end{array}
\longrightarrow
\begin{array}{r} 37.941 \\ 43\,\overline{)\,1631.463} \\ -129 \\ \hline 341 \\ -301 \\ \hline 404 \\ -387 \\ \hline 176 \\ -172 \\ \hline 43 \\ -43 \\ \hline 0 \end{array}
$$

9. 2.175

$$
0.16 \overline{)0.348} \longrightarrow
\begin{array}{r}
2. \\
16 \overline{)34.8} \\
-32 \\
\hline
28
\end{array}
\longrightarrow
\begin{array}{r}
2.1 \\
16 \overline{)34.80} \\
-32 \\
\hline
28 \\
-16 \\
\hline
120
\end{array}
$$

$$
\longrightarrow
\begin{array}{r}
2.17 \\
16 \overline{)34.800} \\
-32 \\
\hline
28 \\
-16 \\
\hline
120 \\
-112 \\
\hline
80
\end{array}
\longrightarrow
\begin{array}{r}
2.175 \\
16 \overline{)34.800} \\
-32 \\
\hline
28 \\
-16 \\
\hline
120 \\
-112 \\
\hline
80 \\
-80 \\
\hline
0
\end{array}
$$

10. 7.652

$$
0.123 \overline{)0.941196} \longrightarrow
\begin{array}{r}
7. \\
123 \overline{)941.196} \\
-861 \\
\hline
801
\end{array}
\longrightarrow
\begin{array}{r}
7.6 \\
123 \overline{)941.196} \\
-861 \\
\hline
801 \\
-738 \\
\hline
639
\end{array}
$$

$$
\longrightarrow
\begin{array}{r}
7.65 \\
123 \overline{)941.196} \\
-861 \\
\hline
801 \\
-738 \\
\hline
639 \\
-615 \\
\hline
24
\end{array}
\longrightarrow
\begin{array}{r}
7.652 \\
123 \overline{)941.196} \\
-861 \\
\hline
801 \\
-738 \\
\hline
639 \\
-615 \\
\hline
246 \\
-246 \\
\hline
0
\end{array}
$$

☞ **Practice: Decimals**

DIRECTIONS: Answer each question as indicated.

1. The place value of 6 in the number 2.1346 is _____.

2. The number 7.012 written in words is _____.

3. The number 205.0098 written in words is _____.

4. For the number 0.3017, the digit in the hundredths place is _____.

5. The number twenty-four and eight hundredths written in digits is _____.

6. The number nine and five hundred twenty ten thousandths is _____.

7. The number 1.6271 when rounded to the hundredths position is _____.

8. The number 0.953 when rounded to the tenths place is _____.

9. The number 0.086464 when rounded to the ten thousandths place is _____.

10. 0.1206 _____ 0.1314 (Fill in the blank with <, =, or >).

11. $13.06 + 1.316 =$ _____.

12. $24.5 + 14.007 + 5.57 =$ _____.

13. $58.109 - 20.773 =$ _____.

14. $4.38 \times 0.075 =$ _____.

15. $0.1108 \times 0.0026 =$ _____.

16. $0.828 \div 23 =$ _____.

17. $11.25 \div 0.15 =$ _____.

18. $12.636 \div 1.8 =$ _____.

19. $96 \div 0.24 =$ _____.

20. $9.9356 \div 4 =$ _____.

21. The place value of 8 in the number 0.817 is _____.

22. For the number 0.6097, the digit in the thousandths place is _____.

23. The number six and fifty-four thousandths written in digits is _____.

24. 0.253 _____ 0.237 (Fill in the blank with <, =, or >.)

25. $1.334 \times 0.028 =$ _____.

Answers

1. Ten thousandths. It is the fourth digit to the right of the decimal point

2. Seven and twelve thousandths.

3. Two hundred five and ninety-eight ten thousandths.

4. 0. This is the second digit to the right of the decimal point.

5. 24.08

6. 9.0520

7. 1.63. Since the thousandths digit is 7, raise the 2 to a 3. Remove the 7 and 1.

8.   1.0.   Since the hundredths digit is 5, raise the 9 to a 0. The invisible 0 to the left of the decimal point becomes a 1. Remove the 5.

9.   0.0865. Since the fifth digit to the right of the decimal point is 6, raise the 4 (fourth digit) to a 5. Remove the two right-most digits.

10.   <   Because the 2 in the hundredths place of the first number is smaller than the 3 in the second number.

11.   14.376

$$\begin{array}{r} 13.06 \\ + 1.316 \\ \hline 14.376 \end{array}$$

12.   44.077

$$\begin{array}{r} \textit{11} \\ 24.5 \\ 14.007 \\ + 5.57 \\ \hline 44.077 \end{array}$$

13.   37.336

$$\begin{array}{r} 58.109 \\ - 20.773 \\ \hline 37.336 \end{array}$$

14.   0.3285 or 0.32850

$$\begin{array}{cccc} 4.38 & & 438 & & 438 \\ \times 0.075 & \rightarrow & \times 0.075 & \rightarrow & \times 0.075 \\ \hline & & 2190 & & 2190 \\ & & & & + 3066 \\ \hline & & & & 32850 \end{array}$$

$$\begin{array}{rl} 4.38 & \text{(2 decimal places)} \\ 0.075 & \text{(3 decimal places)} \\ \hline 2190 & \\ + 3066 & \\ \hline 0.32850 & \text{(5 decimal places)} \end{array}$$

15.   0.00028808

$$\begin{array}{cccc} 0.1108 & & 0.1108 & & 0.1108 \\ \times 0.0026 & \rightarrow & \times 0.0026 & \rightarrow & \times 0.0026 \\ \hline & & 6648 & & 6648 \\ & & & & + 2216 \\ \hline & & & & 28808 \end{array}$$

$$\begin{array}{rl} 0.1108 & \text{(4 decimal places)} \\ \times 0.0026 & \text{(4 decimal places)} \\ \hline 6648 & \\ + 2216 & \\ \hline 0.00028808 & \text{(8 decimal places)} \end{array}$$

16.   0.036

$$\begin{array}{ccc} & 0.03 & 0.036 \\ 23\overline{)0.828} \rightarrow & 23\overline{)0.828} \rightarrow & 23\overline{)0.828} \\ & -69 & -69 \\ \cline{2-3} & 13 & 138 \\ & & -128 \\ \cline{3-3} & & 0 \end{array}$$

17.   75

$$\begin{array}{ccc} & 7 & 75 \\ 0.15\overline{)11.25} \rightarrow & 15\overline{)1125} \rightarrow & 15\overline{)1125} \\ & -105 & -105 \\ \cline{2-3} & 75 & 75 \\ & & -75 \\ \cline{3-3} & & 0 \end{array}$$

18.   7.02

$$\begin{array}{ccc} & 7. & 7.02 \\ 1.8\overline{)12.636} \rightarrow & 18\overline{)126.36} \rightarrow & 18\overline{)126.36} \\ & -126 & -126 \\ \cline{2-3} & 03 & 036 \\ & & -36 \\ \cline{3-3} & & 0 \end{array}$$

19. 400

$$
0.24\overline{)96} \longrightarrow 24\overline{)9600} \longrightarrow 24\overline{)9600}
$$

$$
\begin{array}{r} 4 \\ 24\overline{)9600} \\ -96 \\ \hline 0 \end{array}
\qquad
\begin{array}{r} 400 \\ 24\overline{)9600} \\ -96 \\ \hline 00 \end{array}
$$

20. 2.4839

$$
\begin{array}{r} 2. \\ 4\overline{)9.9356} \\ -8 \\ \hline 1 \end{array}
\longrightarrow
\begin{array}{r} 2.4 \\ 4\overline{)9.9356} \\ -8 \\ \hline 19 \\ -16 \\ \hline 3 \end{array}
\longrightarrow
\begin{array}{r} 2.48 \\ 4\overline{)9.9356} \\ -8 \\ \hline 19 \\ -16 \\ \hline 33 \\ -32 \\ \hline 1 \end{array}
$$

$$
\longrightarrow
\begin{array}{r} 2.483 \\ 4\overline{)9.9356} \\ -8 \\ \hline 19 \\ -16 \\ \hline 33 \\ -32 \\ \hline 15 \\ -12 \\ \hline 3 \end{array}
\longrightarrow
\begin{array}{r} 2.4839 \\ 4\overline{)9.9356} \\ -8 \\ \hline 19 \\ -16 \\ \hline 33 \\ -32 \\ \hline 15 \\ -12 \\ \hline 36 \\ -36 \\ \hline 0 \end{array}
$$

21. Tenths. It is the first digit to the right of the decimal point.

22. 9. The thousandths place is the third digit to the right of the decimal point.

23. 6.054

24. >   Because the hundredths digit of the first number is larger than that of the second number.

25. 0.037352

$$
\begin{array}{r} 1.334 \\ \times\, 0.028 \\ \hline \end{array}
\longrightarrow
\begin{array}{r} \overset{223}{1334} \\ \times\, 0.028 \\ \hline 10672 \end{array}
\longrightarrow
\begin{array}{r} 1334 \\ \times\, 0.028 \\ \hline 10672 \\ +\,2668 \\ \hline 37352 \end{array}
$$

$$
\begin{array}{rl} 1.334 & \text{(3 decimal places)} \\ \longrightarrow \times\, 0.028 & \text{(3 decimal places)} \\ \hline 10672 & \\ +\,2668 & \\ \hline 0.037352 & \text{(6 decimal places)} \end{array}
$$

REVIEW

Learning about decimals can seem frustrating and confusing at first. In dealing with decimals you should remember that all decimals are merely portions of whole numbers. After studying this section, you should be able to identify the place value of digits in decimal numbers, which include the tenths, hundredths, thousandths, ten-thousandths places, etc. You should also be able to convert word statements into decimal numbers and vice versa. These are variations of the skills you learned in the previous section about identifying whole place values.

After reading this section, you should also be able to recognize the different sizes of decimals. You can identify the size of decimals by understanding that when two decimal numbers are nearly the same, the first occurrence of a difference, when moving from the decimal point to the right, will indicate the larger decimal number. For instance, 0.8427 is greater than 0.8413.

When rounding decimal numbers, be sure to check the digit to the right of the desired rounding number; if it is 5 or higher, raise the digit in the desired location by 1, if not, leave that digit alone. For example, 6.81 when rounded to the tenths place, would remain 6.8 because 1 (in the hundredths place) is less than 5. This "rounding rule" is universal, whether you are rounding whole numbers or decimals.

When adding and subtracting decimals, be sure to correctly align the various place holders (tenths, hundredths, etc.). One easy way to ensure that they are properly aligned is to line up the decimal points. For example, when adding the decimal numbers 95.23, 2.64, 4, and 6.479, align them like this:

$$
\begin{array}{r}
95.23 \\
2.64 \\
+\ 4.00 \\
6.479
\end{array}
$$

Multiply decimals the same way that whole numbers are multiplied, and properly place the decimal point in the answer. To do this, add the number of place holders in each number being multiplied.

Divide decimals in a similar fashion to whole numbers. If the divisor is a decimal, move that decimal point until a whole number appears; also adjust the dividend in the same manner. This may require adding additional zeros.

Understanding and working with decimals is something that nearly everyone does everyday and it is important to understand these concepts for many reasons. The ability to add, subtract, divide, and multiply decimals is most commonly used when dealing with money.

Miscellaneous Topics

MATHEMATICS

MISCELLANEOUS TOPICS

AVERAGES

Since we have a good understanding of natural numbers and decimals, we want to move our attention to other topics in mathematics. In this section, we will begin to see how mathematics can be used in physical situations. In order to make our work easier, we first want to discuss the "Order of Operations." These rules give us a recipe, or **algorithm**, to follow when simplifying mathematical expressions. We start at the top of the list and proceed down until all operations have been completed. The table below lists the order of operations. This order makes sure that we all get the same answer from a problem. Most of these ideas have already been discussed, but a few are new to the section. We will explain in detail those topics that have not already been explained.

Order of Operations

1. Do the operations above and below the fraction bar separately.

2. Work inside the parentheses (innermost first). Note that ( ), [ ], and { } can all be used as grouping symbols and are generally referred to as parentheses.

3. Simplify exponents.

4. Multiply and divide from left to right. Note that multiplication and division are done at the same time as we move left to right.

5. Add and subtract from left to right. Note that addition and subtraction are also done

at the same time as we move left to right.

Example 1

Simplify the following expression.

(a) $4 + 2(1 + 6)$

In this example, we must begin inside the parentheses and simplify this part of the expression. Here $1 + 6$ equals 7. If we substitute this value into our expression we get

$4 + 2(7)$

According to our order of operations, the next thing to do is the multiplication (2×7). Even though the multiplication symbol does not appear in the original expression, it is implied. Anytime we have a parenthesis next to another parenthesis or number together without an operation, such as ") (" and "2(", it is understood that multiplication is meant. In this example, 2×7 equals 14. Thus our expression becomes

$4 + 14$

Finally, we can do the addition and get our answer. Here our final answer is 18.

(b) $17 - 3[11 - 2(9 - 6)]$

As with the last example, we must begin inside the parentheses. Since we have a couple of different sets of parentheses, we must work inside the innermost set first. In this example, that would be 9 − 6 or 3. If we substitute this value into our expression, we get

$$17 - 3[11 - 2(3)]$$

Since parentheses still exist in this expression, we need to work inside the remaining parentheses (or brackets). Here, the multiplication 2(3) would be done first since it is ranked higher in our order of operations. At this step, our expression would be

$$17 - 3[11 - 6]$$

Continuing inside the brackets, we get

$$17 - 3[5]$$

As we saw before, multiplication would be handled first. Now we have

$$17 - 15$$

Finally, we can do the subtraction and get our final answer of 2.

(c) $2 + 3\{[13 - 4(1 + 2)] + 5\}$

We must begin inside the innermost parentheses in this example. In this case, 1 + 2 = 3. This gives us

$$2 + 3\{[13 - 4(3)] + 5\}$$

Multiplying inside the innermost parentheses gives us

$$2 + 3\{[13 - 12] + 5\}$$

Next is subtraction inside the brackets. We now have

$$2 + 3\{[1] + 5\} \text{ or } 2 + 3\{1 + 5\}$$

The last operation inside the parentheses is the addition, giving

$$2 + 3\{6\}$$

Next, we want to multiply the 3 and the 6 together to get 18. Substituting this into our expression will give

$$2 + 18$$

Therefore, our final answer is 20.

(d) $\dfrac{1 + 2(7 - 5)}{4(9 - 6) + 3}$

In this example, we must simplify the numerator and denominator of this expression separately. Keep in mind that we must follow the order of operations. In simplifying the numerator, we must work inside the parenthesis first. This gives

$$\frac{1 + 2(2)}{4(9 - 6) + 3}$$

We then do the multiplication in the numerator.

$$\frac{1 + 4}{4(9 - 6) + 3}$$

We now do the addition in the numerator, getting

$$\frac{5}{4(9 - 6) + 3}$$

Next, we follow the same ideas with the denominator.

$$\frac{5}{4(3) + 3}$$

When we do the multiplication in the denominator we get

$$\frac{5}{12 + 3}$$

By doing the addition in the denominator, our expression becomes

$$\frac{5}{15} = 0.\overline{3}$$

Therefore, our answer is 0.3 rounded to one decimal place. The short line appearing above the 3 denotes that the value repeats (0.333333333333…). The expression, 5/15, is a fraction and will be discussed in detail in the next section.

Sometimes finding the middle value of a list of numbers is important to us. Mathematically, this is referred to as the **average**. For example, a number of people take a certain standard test. After the group receives their scores, they want to know the average score. Knowing this, they can compare their scores easily to see how the group did. This can be done by adding the scores together and then dividing by the number of scores (or items). This would be the score of the "typical" student.

Example 2

Find the average of the following lists of numbers.

(a) 75, 81, and 99

In order to calculate this average, we must begin by adding the numbers together. In this case, we get 75 + 81 + 99 = 255. Since we are trying to find the average of three numbers, we divide the sum of the scores by 3. Therefore, 255/3 equals 85. This shows that the "typical" student would score 85. We can also look at this as an equation.

$$\frac{75 + 81 + 99}{3} = \frac{255}{3} = 85$$

(b) 10, 79, 83, 91, 98, and 100

In order to calculate this average, we must begin by adding the numbers

together. In this case, we get 10 + 79 + 83 + 91 + 98 + 100 = 461. Since we are trying to find the average of six numbers, we divide the sum of the scores by 6. Therefore, 461/6 equals 76.83 (rounded to two decimal places). This shows that the "typical" student would score approximately 77. We can also look at this as an equation.

$$\frac{10 + 79 + 83 + 91 + 98 + 100}{6}$$
$$= \frac{461}{6} = 76.83$$

(c) 1, 2, 5, 7, and 9

In order to calculate this average, we must begin by adding the numbers together. In this case, we get 1 + 2 + 5 + 7 + 9 = 24. Since we are trying to find the average of five numbers, we divide the sum of the scores by 5. Therefore, 24/5 equals 4.8. This shows that the "typical" student would score 4.8. We can also look at this as an equation.

$$\frac{1 + 2 + 5 + 7 + 9}{5} = \frac{24}{5} = 4.8$$

(d) 2, 3, 3.25, 3.5, 3.75, 4, and 4

In order to calculate this average, we must begin by adding the numbers together. In this case, we get 2 + 3 + 3.25 + 3.5 + 3.75 + 4 + 4 = 23.5. Since we are trying to find the average of seven numbers, we divide the sum of the scores by 7. Therefore, 23.5/7 equals 3.36 (rounded to two decimal places). This shows that the "typical" student would score approximately 3. We can also look at this as an equation.

$$\frac{2 + 3 + 3.25 + 3.5 + 3.75 + 4 + 4}{7}$$

$$= \frac{23.5}{7} = 3.35714$$

Questions

Simplify the following expressions.

1. $11 + 3(7 - 2)$

2. $40 - 2[5 + 2(12 - 7)]$

3. $9 + \{2[20 - 5(2 + 1)] + 1\}$

4. $\dfrac{2(10 - 3) - 5}{7 - 3(6 - 4)}$

5. $4.7 + 2.1(6.5 - 3.9)$

Find the average of the following numbers.

6. 85, 95, and 97

7. 1, 3, 5, 7, 9, and 9

8. 10, 75, 80, and 92

9. 3, 3.5, 3.5, 3.75, and 4

10. 1, 3, 4, 5, and 5

Answers

1. 26

$$11 + 3(7 - 2)$$
$$= 11 + 3(5)$$
$$= 11 + 15$$
$$= 26$$

2. 10

$$40 - 2[5 + 2(12 - 7)]$$
$$= 40 - 2[5 + 2(5)]$$
$$= 40 - 2[5 + 10]$$
$$= 40 - 2[15]$$
$$= 40 - 30$$
$$= 10$$

3. 20

$$9 + \{2[20 - 5(2 + 1)] + 1\}$$
$$= 9 + \{2[20 - 5(3)] + 1\}$$
$$= 9 + \{2[20 - 15] + 1\}$$
$$= 9 + \{2[5] + 1\}$$
$$= 9 + \{10 + 1\}$$
$$= 9 + \{11\}$$
$$= 20$$

4. 9

$$\frac{2(10 - 3) - 5}{7 - 3(6 - 4)}$$
$$= \frac{2(7) - 5}{7 - 3(2)}$$
$$= \frac{14 - 5}{7 - 6}$$
$$= \frac{9}{1}$$
$$= 9$$

5. 10.16

$$4.7 + 2.1(6.5 - 3.9)$$
$$= 4.7 + 2.1(2.6)$$
$$= 4.7 + 5.46$$
$$= 10.16$$

6. 92.33...

$$
\begin{array}{r}
\overset{1}{85} \\
95 \\
+\,97 \\
\hline
277
\end{array}
\qquad
\begin{array}{r}
92.33 \\
3\,\overline{)277} \\
-\,27 \\
\hline
07 \\
-\,6 \\
\hline
1
\end{array}
$$

7. 5.66...

$$1 + 3 + 5 + 7 + 9 + 9 = 34$$

$$
\begin{array}{r}
5.66 \\
6\,\overline{)34} \\
-\,30 \\
\hline
40 \\
-\,36 \\
\hline
40
\end{array}
$$

8. 64.25

$$
\begin{array}{r}
10 \\
75 \\
80 \\
+\,92 \\
\hline
257
\end{array}
$$

$$
\begin{array}{r}
6 \\
4\,\overline{)257} \\
-\,24 \\
\hline
17
\end{array}
\quad\longrightarrow\quad
\begin{array}{r}
64.25 \\
4\,\overline{)257.00} \\
-\,24 \\
\hline
17 \\
-\,16 \\
\hline
10 \\
-\,8 \\
\hline
20 \\
-\,20 \\
\hline
0
\end{array}
$$

9. 3.55

$$3 + 3.5 + 3.5 + 3.75 + 4 = 17.75$$

$$
\begin{array}{r}
3 \\
5\,\overline{)17.75} \\
-\,15 \\
\hline
2
\end{array}
\longrightarrow
\begin{array}{r}
3.5 \\
5\,\overline{)17.75} \\
-\,15 \\
\hline
27 \\
-\,25 \\
\hline
2
\end{array}
\longrightarrow
\begin{array}{r}
3.55 \\
5\,\overline{)17.75} \\
-\,15 \\
\hline
27 \\
-\,25 \\
\hline
25 \\
-\,25 \\
\hline
0
\end{array}
$$

10. 3.6

$$1 + 3 + 4 + 5 + 5 = 18$$

$$
\begin{array}{r}
3.6 \\
5\,\overline{)18.0} \\
-\,15 \\
\hline
30 \\
-\,30 \\
\hline
0
\end{array}
$$

CUBES AND ROOTS

Previously we mentioned exponents. In this section we will look at exponents and roots. Exponents can be used to denote repetitive multiplication. The following two expressions mean the same thing.

$$11 \times 11 \times 11 \times 11 \quad \text{or} \quad \mathbf{11^4}$$

The second form is the **exponential** form. Since we are multiplying 11 together, we use that value as a **base**. Because we are multiplying four 11's together, we place a 4 in the **exponent position**. This is the little number appearing above and to the right of the base. Using exponents condenses the expression. In the next two sets of examples, we will become familiar with switching to and from exponents.

Example 1

Rewrite the following without using exponents.

(a) 5^2

The base is 5, so we know that we have a line of 5's being multiplied together. Since our exponent is 2, we know that we will have two 5's multiplied; therefore, we get 5×5. The important point here is being able to convert back and forth between the multiplication form and the exponential form. If we want to or are asked to, we could also multiply this out and get 25.

(b) 2^3

The base is 2, so we know that we have a line of 2's being multiplied together. Since our exponent is 3, we know that we will have three 2's multiplied; therefore, we get $2 \times 2 \times 2$. The important point here is being able to convert back and forth between the multiplication form and the exponential form. If we want to or are asked to, we could also multiply this out and get 8.

(c) 7^5

The base is 7, so we know that we have a line of 7's being multiplied together. Since our exponent is 5, we know that we will have five 7's multiplied; therefore, we get $7 \times 7 \times 7 \times 7 \times 7$. The important point here is being able to convert back and forth between the multiplication form and the exponential form. If we want to or are asked to, we could also multiply this out and get 16,807.

(d) 3^4

The base in this case is 3, so we know that we have a line of 3's being multiplied together. Since our exponent is 4, we know that we will have four 3's multiplied; therefore, we get $3 \times 3 \times 3 \times 3$. The important point here is being able to convert back and forth between the multiplication form and the exponential form. If we want to or are asked to, we could also multiply this out and get 81.

Example 2

Rewrite the following using exponents.

(a) $7 \times 7 \times 7$

In this example, we are multiplying a line of 7's together. Therefore, our base is 7. Since we have three 7's being multiplied, our exponent is 3. Thus, we get

7^3

(b) 9×9

In this example, we are multiplying a line of 9's together. Therefore, our base is 9. Since we have two 9's being multiplied, our exponent is 2. Thus, we get

9^2

(c) $2 \times 2 \times 2 \times 2 \times 2$

In this example, we are multiplying a line of 2's together. Therefore, our base is 2. Since we have five 2's being multiplied, our exponent is 5. Thus, we get

2^5

(d) $13 \times 13 \times 13 \times 13$

In this example, we are multiplying a line of 13's together. Therefore, our base is 13. Since we have four 13's being multiplied, our exponent is 4. Thus, we get

13^4

Another related topic is **roots**. We will begin by discussing **square roots**. Later we will see that **cube roots** are very similar. When asked for the square root of a number, we want to find a number that, when we square it, gives us the original number. Knowing this gives us an immediate check to see if our answer is correct. The symbol for square root is something like a check mark ($\sqrt{}$).

Example 3

Find the square root of the following numbers.

(a) $\sqrt{4}$

When asked to find the square root of 4, we want to find a number that when squared equals 4. After some consideration, we can see that our answer is 2. This is because

$2^2 = 2 \times 2 = 4$

and this is the beginning value.

(b) $\sqrt{25}$

When asked to find the square root of 25, we want to find a number that when squared equals 25. After some consideration we can see that our answer is 5. This is because

$5^2 = 5 \times 5 = 25$

and this is the beginning value.

(c) $\sqrt{9}$

When asked to find the square root of 9, we want to find a number that when squared equals 9. After some consideration, we can see that our answer is 3. This is because

$3^2 = 3 \times 3 = 9$

and this is the beginning value.

(d) $\sqrt{100}$

When asked to find the square root of 100, we want to find a number that when squared equals 100. After some consideration, we can see that our answer is 10. This is because

$10^2 = 10 \times 10 = 100$

and this is the beginning value.

The cube root is very similar to the square root, but we are looking for a number that when cubed gives us the original number. The cube root symbol is very similar to the square root symbol with only one exception. We still have a check mark type symbol, but a 3 appears in the upper left corner of the symbol. This is referred to as the **index**.

Example 4

Find the cube root of the following numbers.

(a) $\sqrt[3]{8}$

When asked to find the cube root of 8, we want to find a number that when cubed equals 8. After some consideration, we can see that our answer is 2. This is because

$2^3 = 2 \times 2 \times 2 = 8$

and this is the beginning value.

(b) $\sqrt[3]{125}$

When asked to find the cube root of 125, we want to find a number that when cubed equals 125. After some consideration, we can see that our answer is 5. This is because

$5^3 = 5 \times 5 \times 5 = 125$

and this is the beginning value.

(c) $\sqrt[3]{27}$

When asked to find the cube root of 27, we want to find a number that when cubed equals 27. After some consideration, we can see that our answer is 3. This is because

$3^3 = 3 \times 3 \times 3 = 27$

and this is the beginning value.

(d) $\sqrt[3]{216}$

When asked to find the cube root of 216, we want to find a number that when cubed equals 216. After some consideration, we can see that our answer is 6. This is because

$6^3 = 6 \times 6 \times 6 = 216$

and this is the beginning value.

Now that we have an understanding of exponents, we want to consider what happens when we put them into problems involving order of operations. Recall that the order of operations tells us to simplify the exponents after the parentheses, but before we do the multiplying and dividing. In the following examples, we will show, in detail, the steps needed.

Example 5

Simplify the following expression.

(a) $2 \times 3 + 5 - 2^3$

In this example we need to begin with simplifying the exponent. Two cubed is the same as $2 \times 2 \times 2$ or 8. This converts our expression to

$2 \times 3 + 5 - 8$

Next, we need to simplify the multiplication in our problem. Since 2×3 equals 6, our expression becomes

$6 + 5 - 8$

We now must add and subtract, moving from left to right as they appear. Thus, we add 6 and 5 together first. This gives

$11 - 8$

Finally, we are able to do the subtraction and get 3 as our final answer.

(b) $3^4 - \{50 + 2(6 + 7)\}$

In this example we begin by simplifying inside the parentheses. Working inside the innermost set, we have $6 + 7 = 13$. Our expression is now

$3^4 - \{50 + 2(13)\}$

We can now multiply inside the brackets and get

$3^4 - \{50 + 26\}$

Adding inside the bracket gives

$3^4 - \{76\}$

Now that the parentheses have been dealt with, we can work with the exponent. Since three raised to the fourth is $3 \times 3 \times 3 \times 3$ or 81, our expression becomes

$81 - 76$

Therefore, our final answer is 5.

(c) $7 + \{6 - 2(1 + 1 + 1)\}^{10}$

Working inside the innermost parenthesis, we can begin by adding the ones together. Since $1 + 1 + 1$ equals 3, we can rewrite our expression to get

$7 + \{6 - 2(3)\}^{10}$

Multiplying inside the brackets gives

$7 + \{6 - 6\}^{10}$

If we do the subtraction inside the brackets, we get

$7 + \{0\}^{10}$

The order of operations now tells us that we need to raise zero to the tenth power or $0 \times 0 \times 0 \times 0 \times 0 \times 0 \times 0 \times 0 \times 0 \times 0$. This gives us 0. If we substitute this into our expression, we have

$7 + 0$

Therefore, our final answer is 7.

(d) $9 - 2\{3 + [2 - (9 - 8)^2]\}$

By subtracting inside the innermost parentheses we get

$9 - 2\{3 + [2 - (1)^2]\}$

Now we need to square 1. This gives $1 \times 1 = 1$. Our expression has now become

$9 - 2\{3 + [2 - 1]\}$

By subtracting inside the brackets, we have

$9 - 2\{3 + 1\}$

If we do the addition inside the remaining parentheses we get

$9 - 2\{4\}$

Multiplication is the next thing that we must address. Our expression now becomes

$9 - 8$

Therefore, our final answer is 1.

Questions

Rewrite the following without using exponents.

1. 3^2

2. 7^5

3. 2^4

4. 1^9

5. 4^3

Rewrite the following using exponents.

6. $1 \times 1 \times 1 \times 1 \times 1 \times 1 \times 1$

7. $5 \times 5 \times 5$

8. $6 \times 6 \times 6 \times 6$

9. $2 \times 2 \times 2 \times 2 \times 2$

10. 7×7

Find the square root of the following numbers.

11. 25

12. 49

13 16

14. 81

15. 36

Find the cube root of the following numbers.

16. 1

17. 8

18. 64

19. 125

20. 343

Simplify the following expressions.

21. $(7 \times 2) - 6 - 2^3$

22. $5(3^2 - 7) + 1$

23. $12 - [(4 + 5)^2 - 70]$

24. $3^4 - 4(2^3 + 1) - 8 \times 5$

25. $\dfrac{2(11 - 6) + 5}{5^2}$

Answers

1. 3×3

2. $7 \times 7 \times 7 \times 7 \times 7$

3. $2 \times 2 \times 2 \times 2$

4. $1 \times 1 \times 1 \times 1 \times 1 \times 1 \times 1 \times 1 \times 1$

5. $4 \times 4 \times 4$

6. 1^7

7. 5^3

8. 6^4

9. 2^5

10. 7^2

11. 5

$5 \times 5 = 25$

12. 7

$7 \times 7 = 49$

13. 4

$4 \times 4 = 16$

14. 9

$9 \times 9 = 81$

15. 6

$6 \times 6 = 36$

16. 1

$1 \times 1 \times 1 = 1$

17. 2

$8 = (2 \times 2 \times 2)$

18. 4

$64 = (4 \times 4 \times 4)$

19. 5

$125 = (5 \times 5 \times 5)$

20. 7

$343 = (7 \times 7 \times 7)$

21. 0

$$(7 \times 2) - 6 - 2^3$$
$$= (7 \times 2) - 6 - 8$$
$$= 14 - 6 - 8$$
$$= 8 - 8$$
$$= 0$$

22. 11

$$5(3^2 - 7) + 1$$
$$= 5(9 - 7) + 1$$
$$= 5(2) + 1$$
$$= 10 + 1$$
$$= 11$$

23. 1

$$12 - [(4 + 5)^2 - 70]$$
$$= 12 - [9^2 - 70]$$
$$= 12 - [81 - 70]$$
$$= 12 - [11]$$
$$= 1$$

24. 5

$$3^4 - 4(2^3 + 1) - 8 \times 5$$
$$= 81 - 4(8 + 1) - 8 \times 5$$
$$= 81 - 4(9) - 8 \times 5$$
$$= 81 - 36 - 40$$
$$= 45 - 40$$
$$= 5$$

25. 0.6

$$\frac{2(11 - 6) + 5}{5^2}$$
$$= \frac{2(11 - 6) + 5}{25}$$
$$= \frac{2(5) + 5}{25}$$
$$= \frac{10 + 5}{25}$$
$$= \frac{15}{25}$$
$$= 0.6$$

MEASUREMENT

Working with figures is yet another important part of mathematics. Two common figures that we regularly deal with are **rectangles** and **squares**. A square is a rectangle with all the sides of the same length. As we move through this section, notice that in each problem the units are constant. This is very important when dealing with the units. If the measurements are not already the same, we want to convert them.

We can calculate a few measurements to learn a lot about an object. Some of these mea-

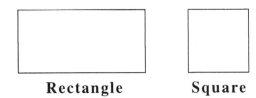

Rectangle   **Square**

surements are perimeter, area, and volume. We will discuss each of these in detail as we move through this section.

The **perimeter** is the distance around the outside of the object. Imagine that we are able to walk around the object and measure the distance we travel. In a rectangle, the perimeter will be twice the length plus twice the width. Notice that if we add numbers together that have inches for units, our answer will also have inches for units. As a formula, we have

length

width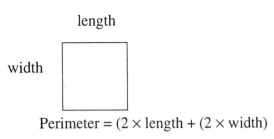

Perimeter = (2 × length + (2 × width)

Example 1

Find the perimeter of a rectangle with the following dimensions.

(a)   length = 5 feet and width = 4 feet

Since we know the length and width of our rectangle, we can substitute these values into our formula. Notice that since our dimensions are given in feet, our answer will also have feet as a unit.

Perimeter = (2 × length) + (2 × width)

Perimeter = (2 × 5) + (2 × 4)

We can now do the multiplication and then the addition to get our answer.

Perimeter = 10 + 8 = 18 feet

(b)   length = 2 inches and width = 1 inch

Since we know the length and width of our rectangle, we can substitute these values into our formula. Notice that since our dimensions are given in inches, our answer will also have inches as a unit.

Perimeter = (2 × length) + (2 × width)

Perimeter = (2 × 2) + (2 × 1)

We can now do the multiplication and then the addition to get our answer.

Perimeter = 4 + 2 = 6 inches

(c)   length = 7 feet and width = 2 feet

Since we know the length and width of our rectangle, we can substitute these values into our formula. Notice that since our dimensions are given in feet, our answer will also have feet as a unit.

Perimeter = (2 × length) + (2 × width)

Perimeter = (2 × 7) + (2 × 2)

We can now do the multiplication and then the addition to get our answer.

Perimeter = 14 + 4 = 18 feet

(d)   length = 9 inches and width = 11 inches

Since we know the length and width of our rectangle, we can substitute these values into our formula. Notice that since our dimensions are given in inches, our answer will also have inches as a unit.

Perimeter = (2 × length) + (2 × width)

Perimeter = (2 × 9) + (2 × 11)

We can now do the multiplication and then the addition to get our answer.

Perimeter = 18 + 22 = 40 inches

We now want to consider what happens to our formula when we look at a square. Since all four sides have the same length, our perimeter will be determined by multiplying the measure of one side and 4 together. In the example below you will see that our work is not much different than for a rectangle. As a formula, this would be

Perimeter = (4 × side)

Example 2

Find the perimeters of the squares with the following measurements on each side.

(a) side = 5 inches

Since we know the measure of each side of our square, we can substitute this value into our formula. Notice that since our dimension is given in inches, our answer will also have inches as a unit.

Perimeter = (4 × side)

Perimeter = (4 × 5)

We can now do the multiplication to get our answer.

Perimeter = 20 inches

(b) side = 1 foot

Since we know the measure of each side of our square, we can substitute this value into our formula. Notice that since our dimension is given in feet, our answer will also have feet as a unit.

Perimeter = (4 × side)

Perimeter = (4 × 1)

We can now do the multiplication to get our answer.

Perimeter = 4 feet

(c) side = 7 inches

Since we know the measure of each side of our square, we can substitute this value into our formula. Notice that since our dimension is given in inches, our answer will also have inches as a unit.

Perimeter = (4 × side)

Perimeter = (4 × 7)

We can now do the multiplication to get our answer.

Perimeter = 28 inches

(d) side = 10 feet

Since we know the measure of each side of our square, we can substitute this value into our formula. Notice that since our dimension is given in feet, our answer will also have feet as a unit.

Perimeter = (4 × side)

Perimeter = (4 × 10)

We can now do the multiplication to get our answer.

Perimeter = 40 feet

Another useful measurement is the **area** of an object. This is the amount of space enclosed by the boundaries of the object. In a rectangle, the area is the length and width multiplied together. Notice that if we multiply two numbers together that have feet as units, our answer will have *square* feet as its unit. As a formula, we have

Area = length × width

Example 3

Find the areas of rectangles with the following dimensions.

(a) length = 2 inches and width = 3 inches

Since we know the length and width of our rectangle, we can substitute these values into our formula. Notice that since our dimensions are given in inches, our answer will have square inches as a unit.

Area = length × width

Area = 2 × 3

We can now do the multiplication to get our answer.

Area = 6 inches²

(b) length = 5 feet and width = 2 feet

Since we know the length and width of our rectangle, we can substitute these values into our formula. Notice that since our dimensions are given in feet, our answer will have square feet as a unit.

Area = length × width

Area = 5 × 2

We can now do the multiplication to get our answer.

Area = 10 feet²

(c) length = 12 inches and width = 1 inch

Since we know the length and width of our rectangle, we can substitute these values into our formula. Notice that since our dimensions are given

in inches, our answer will have square inches as a unit.

Area = length × width

Area = 12 × 1

We can now do the multiplication to get our answer.

Area = 12 inches²

(d) length = 7 feet and width = 4 feet

Since we know the length and width of our rectangle, we can substitute these values into our formula. Notice that since our dimensions are given in feet, our answer will have square feet as a unit.

Area = length × width

Area = 7 × 4

We can now do the multiplication to get our answer.

Area = 28 feet²

When dealing with a square, our area formula becomes a little bit easier. Since all sides have the same length, our formula becomes the measure of one side squared. Written as a formula, we get

Area = side²

Example 4

Find the area of a square with the following measure on each side.

(a) side = 9 inches

Since we know the measure of the sides of our square, we can substitute this value into our formula. Notice that since our dimension is given in inches, our answer will have square inches as a unit.

Area = side^2

Area = $9^2 = 9 \times 9$

We can now simplify this to get our answer.

Area = 81 inches^2

(b) side = 6 feet

Since we know the measure of the sides of our square, we can substitute this value into our formula. Notice that since our dimensions are given in feet, our answer will have square feet as a unit.

Area = side^2

Area = $6^2 = 6 \times 6$

We can now simplify this to get our answer.

Area = 36 feet^2

(c) side = 3 inches

Since we know the measure of the sides of our square, we can substitute this value into our formula. Notice that since our dimension is given in inches, our answer will have square inches as a unit.

Area = side^2

Area = $3^2 = 3 \times 3$

We can now simplify this to get our answer.

Area = 9 inches^2

(d) side = 5 feet

Since we know the measure of the sides of our square, we can substitute this value into our formula. Notice that since our dimensions are given

in feet, our answer will have square feet as a unit.

Area = side^2

Area = $5^2 = 5 \times 5$

We can now simplify this to get our answer.

Area = 25 feet^2

Now we want to consider a box. A useful measure here is the **volume**. The volume is the amount of stuff that can be placed into the object. We can find this measurement by multiplying the length, width, and height together. Notice that if we multiply three numbers together that have inches as their units, our answer will have units of *cubic* inches. As a formula, we have

Volume = length $\times$ width $\times$ height

Example 5

Find the volume of the following box.

(a) length = 6 inches, width = 4 inches, and height = 2 inches

Since we know the measure of the length, width, and height of our box, we can substitute these values into our formula. Notice that since our dimensions are given in inches, our answer will have cubic inches as a unit.

Volume = length $\times$ width $\times$ height

Volume = $6 \times 4 \times 2$

We can now do the multiplication and get our answer.

Volume = $24 \times 2 = 48$ inches^3

(b) length = 1 foot, width = 2 feet, and height = 3 feet

Since we know the measure of the length, width, and height of our box, we can substitute these values into our formula. Notice that since our dimensions are given in feet, our answer will have cubic feet as a unit.

Volume = length × width × height

Volume = 1 × 2 × 3

We can now do the multiplication and get our answer.

Volume = 2 × 3 = 6 feet^3

(c) length = 25 inches, width = 3 inches, and height 6 inches

Since we know the measure of the length, width, and height of our box, we can substitute this value into our formula. Notice that since our dimensions are given in inches, our answer will have cubic inches as a unit.

Volume = length × width × height

Volume = 25 × 3 × 6

We can now do the multiplication and get our answer.

Volume = 75 × 6 = 450 inches^3

(d) length = 10 feet, width = 12 feet, and height = 9 feet

Since we know the measure of the length, width, and height of our box, we can substitute this value into our formula. Notice that since our dimensions are given in feet, our answer will have cubic feet as a unit.

Volume = length × width × height

Volume = 10 × 12 × 9

We can now do the multiplication and get our answer.

Volume = 120 × 9 = 1,080 feet^3

Questions

Find the perimeter of a rectangle with the following dimensions.

1. length = 1 foot and width = 6 feet

2. length = 2 inches and width = 4 inches

3. length = 5 feet and width = 3 feet

4. length = 7 inches and width = 2 inches

5. length = 9 feet and width = 7 feet

Find the perimeter of a square with the following distance on each side.

6. side = 3 inches

7. side = 6 feet

8. side = 2 inches

9. side = 1 foot

10. side = 5 inches

Find the area of a rectangle with the following dimensions.

11. length = 2 inches and width = 5 inches

12. length = 6 feet and width = 1 foot

13. length = 4 inches and width = 7 inches

14. length = 9 feet and width = 12 feet

15. length = 11 inches and width = 6 inches

Find the area of a square with the following distance on each side.

16. side = 11 inches

17. side = 2 feet

18. side = 7 inches

19. side = 12 feet

20. side = 1 inch

Find the volume of the following boxes.

21. length = 4 inches, width = 7 inches, and height = 3 inches

22. length = 5 feet, width = 4 feet, and height = 6 feet

23. length = 17 inches, width = 5 inches, and height = 9 inches

24. length = 2 feet, width = 3 feet, and width = 1 foot

25. length = 6 inches, width = 3 inches, and height = 5 inches

Answers

1. 14 feet

    Perimeter = $(2 \times$ width$) + (2 \times$ length$)$

    $= (2 \times 6) + (2 \times 1)$
    $= 12 + 2 = 14$ feet

2. 12 inches

    $(2 \times$ width$) + (2 \times$ length$)$

    $= (2 \times 4) + (2 \times 2)$

    $= 8 + 4 = 12$ inches

3. 16 feet

    $(2 \times$ width$) + (2 \times$ length$)$

    $= (2 \times 3) + (2 \times 5)$

    $= 6 + 10 = 16$ feet

4. 18 inches

    $(2 \times$ width$) + (2 \times$ length$)$

    $= (2 \times 2) + (2 \times 7)$

    $= 4 + 14 = 18$ inches

5. 32 feet

    $(2 \times$ width$) + (2 \times$ length$)$

    $= (2 \times 7) + (2 \times 9)$

    $= 14 + 18 = 32$ feet

6. 12 inches

    Perimeter = $(4 \times$ side$)$

    $= 4 \times 3 = 12$ inches

7. 24 feet

    Perimeter = $(4 \times$ side$)$

    $= 4 \times 6 = 24$ feet

8. 8 inches

    Perimeter = $(4 \times$ side$)$

    $= 4 \times 2 = 8$ inches

9. 4 feet

    Perimeter = $(4 \times$ side$)$

    $= 4 \times 1 = 4$ feet

10. 20 inches

Perimeter = (4 × side)

= 4 × 5 = 20 inches

11. 10 inches^2

Area = length × width

2 × 5 = 10 inches^2

12. 6 feet^2

Length × width

6 × 1 = 6 feet^2

13. 28 inches^2

Length × width

4 × 7 = 28 inches^2

14. 108 feet^2

Length × width

9 × 12 = 108 feet^2

15. 66 inches^2

Length × width

11 × 6 = 66 inches^2

16. 121 inches^2

Area = side^2

11^2 = 121 inches^2

17. 4 feet^2

Side^2

2^2 = 4 feet^2

18. 49 inches^2

Side^2

7^2 = 49 inches^2

19. 144 feet^2

Side^2

12^2 = 144 feet^2

20. 1 inch^2

Side^2

1^2 = 1 inch^2

21. 84 inches^3

Volume = length × width × height

4 × 7 × 3 = 84 inches^3

4 × 7 = 28

28 × 3 = 84

22. 120 feet^3

Length × width × height

5 × 4 × 6 = 120 feet^3

5 × 4 = 20

20 × 6 = 120

23. 765 inches^3

Length × width × height

17 × 5 × 9 = 765 inches^3

17 × 5 = 85

85 × 9 = 765

24. 6 feet^3

Length × width × height

2 × 3 × 1 = 6 feet^3

2 × 3 = 6

6 × 1 = 6

25. 90 inches^3

Length × width × height

6 × 3 × 5 = 90 inches^3

6 × 3 = 18

18 × 5 = 90

☞ Practice: Topics in Mathematics

DIRECTIONS: Answer each question as indicated.

1. $21 - 2(7 - 3) = $ _____ .

2. $15 + 3[11 + 4(3 - 2)] = $ _____ .

3. $8 + \{5[35 - 4(3 + 1)] - 10\} = $ _____ .

4. $[6 + 3(10 - 2)] \div [4 + 9 \times 4] = $ _____ .

5. The average of 10, 14, 29, and 33 is _____ .

6. The average of 6.4, 4.2, 8.5, 5.3, and 7.1 is _____ .

7. In exponential form, $16 \times 16 \times 16$ can be written _____ .

8. In exponential form, $3 \times 3 \times 3 \times 3 \times 3$ can be written _____ .

9. The value of 4^5 is _____ .

10. The value of $2^7 \times 6^3$ is _____ .

11. The square root of 81 is _____ .

12. The cube root of 125 is _____ .

13. $12 + 3^3 - 10 \times 2 = $ _____ .

14. $40 - 6(8 - 5) + 6^2 = $ _____ .

15. $4^4 - [3(15 - 11) + (1 + 9)^2] = $ _____ .

16. $14 \times 5 - 15 \times 4 \div 5^2 = $ _____ .

17. The perimeter of a rectangle whose length is 13 inches and width is 6 inches = _____ .

18. The perimeter of a square whose side is 5 feet = _____ .

19. The area of a rectangle whose length is 16 inches and width is 7 inches = _____ .

20. The area of a square whose side is 3.2 inches = _____ .

21. The volume of a box with length, width, and height of 22 inches, 5 inches, and 8 inches respectively = _____ .

22. The volume of a box with length, width, and height of 9.6 feet, 4.5 feet, and .6 feet respectively = _____ .

23. The value of $0^{10} \times 1^{10}$ is _____ .

24. The perimeter of a rectangle whose length is 4 inches and width is 2.6 inches = _____ .

25. The area of a square whose side is 5.7 inches = _____ .

Answers

1. 13

$$21 - 2(7 - 3) = 21 - 2(4)$$
$$= 21 - 8 = 13$$

2. 60

$$15 + 3[11 + 4(3 - 2)]$$
$$= 15 + 3[11 + 4\}$$
$$= 15 + 3(15) = 15 + 45 = 60$$

3. 93

$$8 + \{5 [35 - 4 (3 + 1)] - 10\}$$
$$= 8 + \{5 [35 - 16] - 10\}$$
$$= 8 + \{5 [19] - 10\} = 8 + \{85\} = 93$$

4. 0.75

$$[6 + 3 (10 - 2)] \div [4 + 9 \times 4]$$
$$= [6 + 3 (8)] \div [4 + 36]$$
$$= 30 \div 40 = 0.75$$

5. 21.5

$$\frac{10 + 14 + 29 + 33}{4} = \frac{86}{4} = 21.5$$

6. 6.3

$$\frac{6.4 + 4.2 + 8.5 + 5.3 + 7.1}{5} = \frac{31.5}{5} = 6.3$$

7. 16^3

8. 3^5

9. 1,024

$$4^5 = 4 \times 4 \times 4 \times 4 \times 4 = 1,024$$

10. 27,648

$$2^7 = (2 \times 2 \times 2 \times 2 \times 2 \times 2 \times 2) = 128$$
$$6^3 (6 \times 6 \times 6) = 216$$
Then $128 \times 216 = 27,648$

11. 9, since $9 \times 9 = 81$

12. 5, since $5 \times 5 \times 5 = 125$

13. 19

$$12 + 3^3 - 10 \times 2$$
$$= 12 + 27 - 20 = 19$$

14. 58

$$40 - 6 (8 - 5) + 6^2$$
$$= 40 - 6 (3) + 36$$
$$= 40 - 18 + 36 = 58$$

15. 144

$$4^4 - [3 (15 - 11) + (1 + 9)^2]$$
$$= 256 - [3 (4) + 10^2]$$
$$= 256 - [12 + 100]$$
$$= 256 - 112 = 144$$

16. 67.6

$$14 \times 5 - 15 \times 4 \div 5^2$$
$$= 70 - 60 \div 25$$
$$= 70 - 2.4 = 67.6$$

17. 38 inches

Perimeter $= (2 \times 13) + (2 \times 6)$
$$= 26 + 12 = 38 \text{ inches}$$

18. 20 feet

Perimeter $= 4 \times 5 = 20$ feet

19. 112 square inches

Area $= 16 \times 7 = 112$ inches^2

20. 10.24 square inches

Area $= 3.2^2 = 10.24$ inches^2

21. 880 inches^3

Volume $= 22 \times 5 \times 8 = 880$ inches^3

22. 25.92 feet^3

Volume $= 9.6 \times 4.5 \times 0.6 = 25.92$ feet^3

23. 0

$$0^{10} = 0, \ 1^{10} = 1, \ 0 \times 1 = 0$$

24. 13.2 inches

Perimeter $= (2 \times 4) + (2 \times 2.6)$

$= 8 + 5.2 = 13.2$

25. 32.49 inches^2

Area $= 5.7^2 = 32.49$

REVIEW

When using mathematics in physical or "real-life" situations, we need to understand the "Order of Operations." The following list is the standard order of operations:

• Do all operations above and below the fraction bar separately

• Work inside the parentheses first

• Simplify all exponents

• Add, subtract, multiply and divide from left to right—be sure to do them at the same time as you move from left to right.

When finding the average of a group of numbers, be sure to add all the numbers together, then divide the total by the number of numbers originally added. For example, a teacher is calculating average test score grades for a student, and the grades he received were 81, 95, 89, and 78. The teacher would add these grades together for a total of 343, then divide by the number of test scores (in this case 4) to get the average of 85.75 (or, when rounded, 86).

Exponents are used to denote repetitive multiplication. An example of this is 84^3. Essentially, this means $84 \times 84 \times 84$. There are two parts to exponents, the base and the exponent position. In our example cited above, 84 is the base, and three is the exponent position.

A similar function is used when dealing with roots. There are two common types of roots, square roots and cube roots. A square root of a number yields a number, which, when multiplied by itself, will return to the original number. (For instance, five squared is 25: $5 \times 5 = 25$.) A cube root yields a number that, when cubed, returns to the original number. (For instance, two cubed equals eight: $2 \times 2 \times 2 = 8$.)

Another important part of practical mathematics is working with rectangles and squares. Remember that the perimeter (or distance around the outside) of a rectangle equals twice the length added to twice the width. The perimeter of a square equals four times the length of one side. Area is the total amount of space enclosed by the boundaries of an object. To determine the area of a rectangle, multiply the length times the width. The area of a square equals one side squared. By multiplying the length, times width, times height of an object, you can determine the total volume of that object.

It is common to need to know how to find the perimeters and areas of squares and rectangles when building a fence around the perimeter of your yard, laying carpeting, or determining the amount of paint or wallpaper to purchase. These are just some of the practical ways that mathematics can be used in everyday life.

Mathematics

Percents

MATHEMATICS

PERCENTS

INTRODUCTION TO PERCENTS

Percents play an important role in our lives. It is hard to pick up a newspaper or listen to the news without hearing the word percent or seeing the symbol (%). Some examples of this are: "Crime Rate Up 32%," "Car Prices Increase 10%," and "Going Out of Business Sale—75% Off of Everything."

The easiest way to understand percents is to think of them as so many out of 100. If, at a 100-person party, 15 of the people are males, then we know that 15% of the people are male. Likewise, if 57% of the city voted for a school levy, then we know that out of every 100 people, 57 voted yes. If the city's population is 200 people, then 114 people voted for the levy. As we saw in the last example, we do not

always have 100 people, but when working with percents it is just easier to think in that format.

If we are given information about a situation where exactly 100 people are not involved, we can convert it to a percent by dividing the number of people that meet some type of requirement by the total group. This "answer" will come out as a decimal. In order to write this value using the percent symbol, we must move the decimal point two places to the *right*. Therefore, for example, 0.39 would represent 39%. Thus, out of every 100 people, 39 would act in a certain way.

Example 1

Write the following decimal using the percent symbol.

(a) 0.24

In order to change this into a percent, we must move the decimal point two places to the right. Therefore, 0.24 would be 24%.

(b) 0.457

In order to change this into a percent, we must move the decimal point two places to the right. Therefore, 0.457 would be 45.7%.

(c) 1.52

In order to change this into a percent,

we must move the decimal point two places to the right. Therefore, 1.52 would be 152%.

(d) 0.9

In order to change this into a percent, we must move the decimal point two places to the right. Since this example does not have two places to the right of the decimal point, we can use a zero as a place holder and re-write our problem as 0.90. Therefore, 0.90 would be 90%.

It is also helpful to be able to convert percentages to decimals. In order to change a percent into its decimal value, we must move the decimal point two places to the *left*. Therefore, 16% is the same as 0.16. As we will see in the next section, when working with a percent, we must remember to convert to a decimal before any calculations are done.

Example 2

Write the following percent as a decimal.

(a) 38%

In order to convert a percent into a decimal, we must move the decimal point two places to the left. Even though the original problem does not contain a decimal point, one is assumed to be after the last digit. This can be written as 38.%. Therefore, 38% would be 0.38.

(b) 61.3%

In order to convert a percent into a decimal, we must move the decimal point two places to the left. Therefore, 61.3% would be 0.613.

(c) 7%

In order to convert a percent into a decimal, we must move the decimal point two places to the left. Since we only have one digit here, it might to a good idea to place a zero in front of the seven so that we can see two positions. This gives us 07%. Therefore, 7% would be 0.07.

(d) 378%

In order to convert a percent into a decimal, we must move the decimal point two places to the left. Therefore, 378% would be 3.78.

As we mentioned before, we can find the percent of something by dividing. Another way to look at this is in a fractional form. The numerator contains the number of people who meet a certain requirement. The denominator contains the total number of people. In the examples below, we will see how easy it is to find the percent.

Example 3

Find the percent in the following situation.

(a) 6 out of 8

In order to find the percent here, we place the 6 in the numerator and the 8 in the denominator. This division gives us 6/8 or 0.75. From our discussion above, we know that this is 75%.

(b) 9 out of 72

In order to find the percent here, we place the 9 in the numerator and the 72 in the denominator. This division gives us 9/72 or 0.125. From our discussion above, we know that this is 12.5%.

(c)   7 out of 10

In order to find the percent here, we place the 7 in the numerator and the 10 in the denominator. This division gives us 7/10 or 0.7. From our discussion we know that this is 70%.

(d)   4 out of 28

In order to find the percent here, we place the 4 in the numerator and the 28 in the denominator. This division gives us 4/28 or 0.1428571. From our discussion we know that this is 14.29%, rounded to two decimal places.

Questions

Write the following decimals using the percent symbol.

1.   0.59

2.   0.2341

3.   7.04

4.   0.085

5.   1.92154

Write the following percents as decimals.

6.   38%

7.   17.4%

8.   5.2%

9.   67.258%

10.   49.1%

Find the percents in the following situations.

11.   5 out of 25

12.   2 out of 20

13.   45 out of 405

14.   28 out of 42

15.   101 out of 191

Answers

1.   59%

Move the decimal point two places to the right to convert from a decimal to a percent.

2.   23.41%

Move the decimal point two places to the right to convert from a decimal to a percent.

3.   704%

Move the decimal point two places to the right to convert from a decimal to a percent.

4.   8.5%

Move the decimal point two places to the right to convert from a decimal to a percent.

5.   192.154%

Move the decimal point two places to the right to convert from a decimal to a percent.

6.   0.38

Move the decimal point two places to the left when converting a percent to a decimal.

7.   0.174

Move the decimal point two places to the left when converting a percent to a decimal.

8. 0.052

Move the decimal point two places to the left when converting a percent to a decimal.

9. 0.67258

Move the decimal point two places to the left when converting a percent to a decimal.

10. 0.491

11. 20%

$$5 \text{ out of } 25 = \frac{5}{25} = 0.2 = 20\%$$

12. 10%

$$2 \text{ out of } 20 = \frac{2}{20} = 0.1 = 10\%$$

13. 11.11...%

$$45 \text{ out of } 405 = \frac{45}{405}$$

$$= 0.1111...$$

$$= 11.11...\%$$

14. 66.66...%

$$28 \text{ out of } 42 = \frac{28}{42}$$

$$= 0.6666...$$

$$= 66.66...\%$$

15. 52.879581%

$$101 \text{ out of } 191 = \frac{101}{191}$$

$$= 0.52879581$$

$$= 52.879581...\%$$

PERCENT PROBLEMS—WORKING WITH PERCENTS

Now that we have an introduction to percents, we want to use them in problems. We have already discussed how to denote percents and convert them from decimals to percents, and how to find out what percent a number is of another number. Here we want to consider the other possibilities. In the example below, we will see how to find a certain percent of a number, and how to solve increase and decrease problems. A basic formula for percents is:

Amount = Percent × Beginning

In most percent problems this formula can be manipulated into a form that will help us.

Example 1

Simplify the following percent problems.

(a) 16% of 30

Here we are given the percent and the beginning amount. We are asked to find the final amount. When working with problems involving words, "of" normally means multiplication. Also recall that when working with percents, we need to convert them into decimals. Therefore, this problem becomes

16% of 30 or 0.16 × 30

We are then able to do the multiplication and simplify this to 4.8. Thus, 4.8 is 16% of 30.

(b) 73% of 157

Here we are given the percent and the beginning amount. We are asked to find the final amount. When working with problems involving words, "of" normally means multiplication. Also recall that when working with percents, we need to convert them into decimals. Therefore, this problem becomes

73% of 157 or 0.73 × 157

We are then able to do the multiplication and simplify this to 114.61. Thus, 114.61 is 73% of 157. If we round our answer to one decimal place we get 114.6.

(c) 248% of 500

Here we are given the percent and the beginning amount. We are asked to find the final amount. When working with problems involving words, "of" normally means multiplication. Also recall that when working with percents, we need to convert them into decimals. Therefore, this problem becomes

248% of 500 or 2.48 × 500

We are then able to do the multiplication and simplify this to 1,240. Thus, 1,240 is 248% of 500.

(d) 51.9% of 6

Here we are given the percent and the beginning amount. We are asked to find the final amount. When working with problems involving words, "of" normally means multiplication. Also recall that when working with percents, we need to convert them into decimals. Therefore, this problem becomes

51.9% of 6 or 0.519 × 6

We are then able to do the multiplication and simplify this to 3.114. Thus, 3.114 is 51.9% of 6. If we round our answer to one decimal place we get 3.1.

Example 2

Find the numbers described below.

(a) 25% of some number is 12. Find the number.

In this problem, we are given the percent and the final amount, but not the beginning amount. We also know that we must change the percent to a decimal in order to work with it. To find our answer, we must divide the final amount by the percent. Here we have

$$\frac{12}{25\%} \text{ or } \frac{12}{0.25}$$

When we divide, we get 48. This tells us that 25% of 48 is 12.

(b) 17% of some number is 51. Find the number.

In this problem, we are given the percent and the final amount, but not the beginning amount. We also know that we must change the percent to a decimal in order to work with it. To find our answer, we must divide the final amount by the percent. Here we have

$$\frac{51}{17\%} \text{ or } \frac{51}{0.17}$$

When we divide, we get 300. This tells us that 17% of 300 is 51.

(c) 85.6% of some number is 305.592. Find the number.

In this problem, we are given the percent and the final amount, but not the beginning amount. We also know that we must change the percent to a decimal in order to work with it. To find our answer, we must divide the final amount by the percent. Here we have

$$\frac{305.592}{85.6\%} \text{ or } \frac{305.592}{0.856}$$

When we divide, we get 357. This tells us that 85.6% of 357 is 305.592.

(d)   112% of some number is 56. Find the number.

In this problem, we are given the percent and the final amount, but not the beginning amount. We also know that we must change the percent to a decimal in order to work with it. To find our answer, we must divide the final amount by the percent. Here we have

$$\frac{56}{112\%} \text{ or } \frac{56}{1.12}$$

When we divide, we get 50. This tells us that 112% of 50 is 56.

Now that we have considered problems involving percents in different ways, we want to look at interest problems. Most of these types of situations come from business and banking. The main formula that we will use here is

Interest = Principal × Rate × Time

In this formula, the "Principal" is the amount that we originally placed into the account, the "Rate" is the interest rate the account is earning in a decimal form, and the "Time" is how long we leave the money in the account measured in years.

Example 3

In the following problems, find the amount of interest earned.

(a) $500 is placed into a saving account paying 3.5% for two years. How much interest will be earned?

In this problem, we know the Principal, Rate, and Time. We can plug these values into our formula and calculate the amount of interest that we would earn.

Interest = Principal × Rate × Time

Interest = $500 × 3.5% × 2 years

Interest = 500 × 0.035 × 2

Interest = $35

Therefore, if we place $500 into a saving account paying 3.5% for two years, we would earn $35 in interest.

(b)   A mutual fund has earned 47% for the last six years. If we had placed $1,200 in this account six years ago, how much interest would we have earned?

In this problem, we know the Principal, Rate, and Time. We can plug these values into our formula and calculate the amount of interest that we would earn.

Interest = Principal × Rate × Time

Interest = $1,200 × 47% × 6 years

Interest = 1,200 × 0.47 × 6

Interest = $3,384

Therefore, if we had placed $1,200 into this mutual fund account paying

47% for six years, we would earn $3,384 in interest.

(c) We place $25,000 in a money market account for 10 years. If the account pays 7%, how much interest will we earn?

In this problem, we know the Principal, Rate, and Time. We can plug these values into our formula and calculate the amount of interest that we would earn.

Interest = Principal × Rate × Time

Interest = $25,000 × 7% × 10 years

Interest = 25,000 × 0.07 × 10

Interest = $17,500

Therefore, if we place $25,000 into a money market account paying 7% for 10 years, we would earn $17,500 in interest.

(d) Historically, a stock has earned 21%. If we place $200 in this stock for 1½ years, how much interest would be earned?

In this problem we know the Principal, Rate, and Time. We can plug these values into our formula and calculate the amount of interest that we would earn.

Interest = Principal × Rate × Time

Interest = $200 × 21% × 1½ years

Interest = 200 × 0.21 × 1.5

Interest = $63

Therefore, if we place $200 into a stock paying 21% for 1½ years, we would earn $63 in interest.

Questions

Simplify the following percent problems.

1. 31% of 50

2. 77% of 294

3. 59% of 600

4. 5.64% of 235

5. 128% of 474

Find the numbers described below.

6. 12% of some number is 25. Find the number.

7. 67% of some number is 46. Find the number.

8. 89% of some number is 512. Find the number.

9. 31.55% of some number is 64. Find the number.

10. 310% of some number is 452. Find the number.

In the following problems, find the amount of interest earned.

11. $250 is placed into a checking account paying 2.7%. If no other money is deposited and no money is withdrawn, how much interest would be earned after five years?

12. If we place $15,000 into a certificate of deposit (CD) paying 12%, how much interest would be earned after seven years?

13. A certain bond account normally earns 37%. If we had placed $27,501 into this

account 25 years ago, how much interest would have been earned?

14. A mutual fund has earned 49% for the last two years. If we had placed $5,000 into this account, how much interest would we have earned?

15. The stock of the KAN Company has earned approximately 20.5% for the last ten years. If we had placed $1,234 into this stock nine years ago, how much interest would have been earned?

Answers

1. 15.5

$$31\% \text{ of } 50 = 31\% \times 50 = 0.31 \times 50$$
$$= 15.5$$

2. 226.38

$$77\% \text{ of } 294 = 77\% \times 294$$
$$= 0.77 \times 294 = 226.38$$

3. 354

$$59\% \text{ of } 600 = 59\% \times 600$$
$$= 0.59 \times 600 = 354$$

4. 13.254

$$5.64\% \text{ of } 235 = 5.64\% \times 235$$
$$= 0.0564 \times 235 = 13.254$$

5. 606.72

$$128\% \text{ of } 474 = 128\% \times 474$$
$$= 1.28 \times 474 = 606.72$$

6. 208.33...

The required number is

$$25 \div 12\% = 25 \div 0.12 = 208.33...$$

7. 68.656716

The required number is

$$46 \div 67\% = 46 \div 0.67 = 68.656716$$

8. 575.2809

The required number is

$$512 \div 89\% = 512 \div 0.89 = 575.2809$$

9. 202.85261

The required number is

$$64 \div 31.55\% = 64 \div 0.3155$$
$$= 202.85261$$

10. 145.80645

The required number is

$$452 \div 310\% = 452 \div 3.10$$
$$= 145.80645$$

11. $33.75

$$\text{Interest} = \$250 \times 2.7\% \times 5$$
$$= \$250 \times 0.027 \times 5 = \$33.75$$

12. $12,600

$$\text{Interest} = \$15,000 \times 12\% \times 7$$
$$= \$15,000 \times 0.12 \times 7 = \$12,600$$

13. $254,384.25

$$\text{Interest} = \$27,501 \times 37\% \times 25$$
$$= \$27,501 \times 0.37 \times 25$$
$$= \$254,384.25$$

14. $4,900

$$\text{Interest} = \$5,000 \times 49\% \times 2$$
$$= \$5,000 \times 0.49 \times 2 = \$4,900$$

15. $2,276.73

$$\text{Interest} = \$1,234 \times 20.5\% \times 9$$
$$= \$1,234 \times 0.205 \times 9 = \$2,276.73$$

☞ **Practice: Percents**

DIRECTIONS: Answer the following questions as indicated.

1. 0.059 written as a percent is _____.

2. 3.7 written as a percent is _____.

3. 0.4% written as a decimal is _____.

4. 221% written as a decimal is _____.

5. 9 out of 12 represents _____%.

6. 10 out of 16 represents _____%.

7. 17% of 25 = _____.

8. 28.3% of 8 = _____.

9. 132% of 12.5 = _____.

10. 24% of the number _____ is 7.2.

11. 6.4% of the number _____ is 0.352.

12. 250% of the number _____ is 35.

13. .3% of the number _____ is 9.75.

14. 540% of the number _____ is 62.1.

15. $1,000 is placed into a savings account paying 4.5% for three years. The interest earned is _____.

16. $8,500 is placed into a money market account for six years, and the account pays 8%. The interest earned is _____.

17. $675 is placed into a stock for $7\frac{1}{2}$ years, and the stock earns 22%. The interest earned is _____.

18. $2,400 is placed into a mutual fund paying 32.5% for $1\frac{1}{2}$ years. The interest earned is _____.

19. $440 is placed into a bond account paying 18.2% for four years. The interest earned is _____.

20. $1,800 is placed into a checking account paying 3.45% for five years. The interest earned is _____.

21. 0.77% written as a decimal is _____.

22. 15 out of 35 represents _____ %.

23. 2.6% of 20 = _____.

24. 18% of the number _____ is 13.68.

25. 420% of the number _____ is 67.2.

Answers

1. 5.9%

   Move the decimal point two places to the right.

2. 370%

   Move the decimal point two places to the right and add one zero.

3. 0.004

   Move the decimal point two places to the left and add two zeros.

4. 2.21

   Move the decimal point two places to the left.

5. 75%

$$\frac{9}{12} = 0.75 = 75\%$$

6. 62.5%

$$\frac{10}{16} = 0.625 = 62.5\%$$

7. 4.25

$$0.17 \times 25 = 4.25$$

8. 2.264

$$0.283 \times 8 = 2.264$$

9. 16.5

$$1.32 \times 12.5 = 16.5$$

10. 30

$$7.2 \div 24\% = 7.2 \div 0.24 = 30$$

11. 5.5

$$0.352 \div 6.4\% = 0.352 \div 0.064 = 5.5$$

12. 14

$$35 \div 250\% = 35 \div 2.50 = 14$$

13. 3,250

$$9.75 \div 0.3\% = 9.75 \div 0.003 = 3,250$$

14. 11.5

$$62.1 \div 540\% = 62.1 \div 5.40 = 11.5$$

15. $135

$$\$1,000 \times 0.045 \times 3 = \$135$$

16. $4,080

$$\$8,500 \times 0.08 \times 6 = \$4,080$$

17. $1,113.75

$$\$675 \times 0.22 \times 7\frac{1}{2} = \$1,113.75$$

18. $1,170

$$\$2,400 \times 0.325 \times 1\frac{1}{2} = \$1,170$$

19. $320.32

$$\$440 \times 0.182 \times 4 = \$320.32$$

20. $310.50

$$\$1,800 \times 0.0345 \times 5 = \$310.50$$

21. 0.0077

Move the decimal point to the left two places and add two zeros.

22. 42.86%

$$\frac{15}{35} \approx 0.4286 = 42.86\%$$

(Note: $\approx$ means approximate.)

23. 0.52

$$0.026 \times 20 = 0.52$$

24. 76

$$13.68 \div 0.18 = 76$$

25. 16

$$67.2 \div 4.20 = 16$$

REVIEW

Percentages are commonly found in our everyday lives. In many cases we often hear of prices of merchandise going up five percent, or that our interest rate at the bank is eight percent. Similarly, we see advertisements for end of the year sales, where everything is 75% off, or that employees are given a ten percent discount. The easiest way to understand the concept of percent is to know that it means "so many out of 100." For example, if we have 100 books and 59 were damaged in a flood, we can say that 59% of the books are damaged.

Most of the time, we do not conveniently deal with situations where the total number is one hundred. In these cases, some conversion is necessary. You can convert a number to a percent by dividing the number of people (or things) that meet the requirement by the total group. This number will be a decimal. At any time, to change from a decimal number to a percent, you need to move the decimal point two places to the right and add the percent sign (%). For instance 0.48 (forty-eight hundredths, or 48/100) is the same as 48%.

To change from a fraction to a percent, divide the denominator into the numerator to obtain the decimal. Then, move the decimal point two places to the right and attach the percent sign. For example, to find 15% of 32, calculate $0.15 \times 32 = 4.80$. If we are given that 20% of some number is 10, calculate $10/20\% = 10 \div 0.20 = 50$.

When dealing with interest and interest problems, remember that interest equals Principal times rate times time, or in an equation form: Interest = Principal $\times$ Rate $\times$ Time. This is considered the formula for simple interest.

Mathematics

Algebra

MATHEMATICS

ALGEBRA

INTRODUCTION TO ALGEBRA

In algebra, letters of variables are used to represent numbers. A **variable** is defined as a placeholder, which can take on any of several values at a given time. A **constant**, on the other hand, is a symbol which takes on only one value at a given time. A **term** is a constant, a variable, or a combination of constants and variables. For example: 7.76, $3x$, xyz, $5z/x$, $(0.99)x^2$ are terms. If a term is a combination of constants and variables, the constant part of the term is referred to as the **coefficient** of the variable.

If a variable is written without a coefficient, the coefficient is assumed to be 1. For example in the term $3x^2$, the coefficient is 3 and the variable is x. Likewise, in the term y^3, the coefficient is 1 and the variable is y.

An **expression** is a collection of one or more terms. If the number of items is greater than 1, the expression is said to be the sum of the terms. Some examples are 9, $9xy$, $6x + x/3$, and $8yz - 2x$. Note that when no operation sign is shown between terms in an expression, multiplication is implied. For example, the expression xyz would be $x \times y \times z$.

An algebraic expression consisting of only one term is called a **monomial**; of two terms is called a **binomial**; of three terms is called a **trinomial**. In general, an algebraic expression consisting of two or more items is called a **polynomial**.

Example 1

Identify the parts of the following expressions.

(a) $2x^5$

In this expression, there is one term with two parts. There is a coefficient, 2, and a variable, x.

(b) $8xz^2$

In this expression, there is one term with three parts. There is a coefficient, 8, a variable x, and another variable z.

(c) $9x + y$

In this expression, there are two terms. The first term has two parts, a coefficient, 9, and a variable, x; and the second term has only a variable, y.

Example 2

Specify whether each of the following expressions is a monomial, binomial, or trinomial.

(a) $\dfrac{5y^5}{z^2}$

Although this expression has several parts, it has only one term, $5y^5/z^2$. Therefore, this expression is a monomial.

(b) $15 + y^3 - \dfrac{8}{z}$

This expression has three terms, 15*x*, *y*³, and 8/*z*. Therefore, this expression is a trinomial.

(c) 4*xyz*⁶

Although this expression has several parts, it has only one term, 4*xyz*⁶. Therefore, this expression is a monomial.

(d) 3*r*³ – 2*s*²

This expression has two parts, 3*r*³ and 2*s*². Therefore, this expression is a binomial.

INTEGERS AND REAL NUMBERS

In mathematics, there are times when it is useful to categorize numbers in groups called **sets**. Most of the numbers used in algebra belong to a set called the **real numbers** or **reals**. This set can be represented graphically by the real number line.

Given the number line below we arbitrarily fix a point and label it with the number 0. In a similar manner, we can label any point on the line with one of the real numbers, depending on its position relative to 0. Numbers to the right of zero are positive, while those to the left are negative. Value increases from left to right, so that if *a* is to the right of *b*, it is said to be greater than *b*.

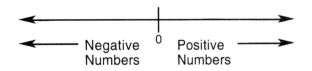

If we now divide the number line into equal segments, we can label the points on this line with real numbers. For example, the point 2 lengths to the left of zero is –2, while the point 3 lengths to the right of zero is + 3 (the + sign is usually assumed, so + 3 is written simply as 3). The number line now looks like this:

These boundary points represent a **subset**, or a group within a set, of the reals known as the **integers**. The set of integers is made up of both the positive and negative whole numbers: {… –4, –3, –2, –1, 0, 1, 2, 3, 4, …}. Some subsets of integers follow.

Subsets of Integers

1. **Natural Numbers** or **Positive Numbers** are the set of integers starting with 1 and increasing: N = {1, 2, 3, 4, …}.

2. **Whole Numbers** are, as we learned earlier, the set of integers starting with 0 and increasing: W = {0, 1, 2, 3, …}.

3. **Negative Numbers** are the set of integers starting with –1 and decreasing: Z = {–1, –2, –3 …}.

4. **Prime Numbers** are the set of positive integers greater than 1 that are divisible only by 1 and themselves: {2, 3, 5, 7, 11, …}.

5. **Even Integers** are the set of integers divisible by 2: {…, –4, –2, 0, 2, 4, 6, …}.

6. **Odd Integers** are the set of integers not divisible by 2: {…, –3, –1, 1, 3, 5, 7, …}.

Example 1

Classify each of the following numbers into as many different sets as possible. Example: real, integer …

(a) 0

Zero is a real number, an integer, and a whole number.

(b) 9

9 is a real number, an integer, an odd number, and a natural number.

(c) $\sqrt{6}$

$\sqrt{6}$ is a real number.

(d) $^1/_2$

$^1/_2$ is a real number.

(e) $^2/_3$

$^2/_3$ is a real number.

(f) 1.5

1.5 is a real number, and a decimal.

Absolute Value

The **absolute value** of a number is represented by two vertical lines around the number, and is equal to the given number, regardless of sign.

The absolute value of a real number A is defined as follows:

$$|A| = \begin{cases} A & \text{if } A \geq 0 \\ -A & \text{if } A < 0 \end{cases}$$

We can think of a minus sign (–) before a constant or variable that indicates the negative as the "opposite" of positive. So if one reads –(–4), this indicates "the opposite of –4," or "the opposite of the opposite of 4," which is 4.

Notice the new symbol, $\geq$. It looks like a cross between > and = because that is exactly how it is used. It is read, "greater than or equal to." As you might guess, there is also a symbol, $\leq$, which is used to mean "lesser than or equal to."

To better understand absolute values, we need to discuss operational functions involving negatives. In order to add two numbers with like signs, add their absolute values and write the sum with the common sign, So,

$6 + 2 = 8$

$(-6) + (-2) = -8$

To add two numbers with unlike signs, find the difference between their absolute values, and write the result with the sign of the number with the greater absolute value. So,

$(-4) + 6 = 6 - 4 = 2$

$15 + (-19) = 15 - 19 = -4$

In the case of adding two negatives, we see that adding a negative number is the same as subtracting a positive number. But we need to be careful to properly place the positive and negative signs, to avoid confusion.

Example 2

Find the absolute value for each of the following questions.

(a) 5

Since 5 is already positive, $|5| = 5$.

(b) –8

Since –8 is negative, we need to get the opposite of –8. Thus, $|-8| = -(-8) = 8$.

(c) 17

Since 17 is already positive, $|17| = 17$.

(d) –x

Since –x is negative, we need to get the opposite of –x. Thus, $|-x| = -(-x) = x$.

To multiply (or divide) two numbers having like signs, multiply (or divide) their absolute values and write the result with a positive sign. Examples:

$(5)(3) = 15$

$(-6) \div (-3) = 2$

To multiply (or divide) two numbers having unlike signs, multiply (or divide) their absolute values and write the result with a negative sign. Examples:

$$(-2)(8) = -16$$

$$9 \div (-3) = -3$$

According to the law of signs for real numbers, the square of a positive or negative number is always positive. This means that it is impossible to take the square root of a negative number in the real number system.

Example 3

Classify each of the following statements as true or false.

(a) $|-A| = -|A|$

False. The absolute value of $|A| = A$; likewise, $|-A| = A$. So $|-A| = |A|$.

(b) $|A| > 0$

False. $|A|$ is not greater than 0 when $A = 0$.

(c) $\left|\dfrac{A}{B}\right| = \dfrac{|A|}{|B|}, B \neq 0$

True. Regardless of whether A and B are positive or negative, after dividing one by the other, the absolute value of the answer will be positive, as will division of the absolute values of A and B.

(d) $|AB| = |A| \times |B|$

True. Regardless of whether A and B are positive or negative, after multiplying one by the other, the absolute value of the answer will be positive, as will multiplication of the absolute values of A and B.

(e) $|A|^2 = |-A|^2 = A^2$

True. Since multiplying two numbers with like signs always yields a positive answer, and we know $|A|^2 = |A| \times |A|$ (and $|-A|^2 = |-A| \times |-A|$),

squaring a number will always yield a positive result.

Absolute value can also be expressed on the real number line as the distance of the point represented by the real number from the point labeled 0.

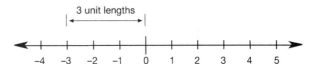

So $|-3| = 3$ because -3 is 3 units to the left of 0.

Example 4

Classify each of the following statements as true or false.

(a) $|-120| > 1$

True. Since -120 is negative, we need to get the opposite of -120. Thus, $|-120| = -(-120) = 120$, and 120 is greater than 1.

(b) $|4 - 12| = |4| - |12|$

False. Keep in mind the proper order of operations. For $|4 - 12|$, subtract first, and then take the absolute value of the difference. So $|4 - 12| = |-8| = 8$. For $|4| - |12|$, however, we find the absolute values of each term before subtracting, so $|4| - |12| = 4 - 12 = -8$. $8 \neq -8$.

(c) $|4 - 9| = 9 - 4$

True. Keep in mind the proper order of operations. For $|4 - 9|$, subtract first, and then take the absolute value of the difference. So $|4 - 9| = |-5| = 5$. Since $9 - 4 = 5$, we see that $5 = 5$.

(d) $|12 - 3| = 12 - 3$

True. Keep in mind the proper order of operations. For $|12 - 3|$, subtract first, and then take the absolute value of the difference. So $|12-3| = |9| = 9$. Since $12 - 3 = 9$, we see that $9 = 9$.

(e) $|-12a| = 12|a|$

True. Keep in mind the proper order of operations. For $|-12a|$, multiply first, and then take the absolute value of the result. So $|-12a| = 12a$. For $12|a|$, however, we find the absolute value of a before multiplying, so $12|a| = 12a$. $12a = 12a$.

Example 5

Calculate the value of each of the following expressions.

(a) $||2-5|+6-14|$

Keep in mind the proper order of operations. Subtract $|2-5|$ first, as you would if parentheses were used instead of the vertical absolute value lines, so that we get $||-3|+6-14|$. Next solve $|-3|$, to give us $|3+6-14| = |9-14| = |-5| = 5$.

(b) $|-5| \times |4| + \dfrac{|-12|}{4}$

Keep in mind the proper order of operations. Solve the absolute values of $|-5|$, $|4|$, and $|-12|$ first, as you would if parentheses were used instead of the vertical absolute value lines, so that we get $(5)(4) + {}^{12}/_4 = (5 \times 4) + (12 \div 4) = 20 + 3 = 23$.

(c) $\dfrac{|12/3|}{|-2^2|}$

Keep in mind the proper order of operations. First divide $12 \div 3 = 4$ and solve $(-2)^2 = 4$, as you would if parentheses were used instead of the vertical absolute value lines, so that

we get $\dfrac{|4|}{|4|} = 1$.

(d) $\left|\dfrac{6-17}{-3+8}\right| \times \sqrt{|(-13+7)+(-19)|}$

Keep in mind the proper order of operations. Since the absolute value of a fraction is equal to the absolute value of the numerator divided by the absolute value of the denominator, start by subtracting $6 - 17 = -11$, and adding $-3 + 8 = 5$ and $(-13 + 7) = -6$, which gives us $\left|\dfrac{-11}{5}\right| \times \sqrt{|(-6)+(-19)|}$. Next add $(-6) + (-19) = -25$. Then, taking the absolute values where marked, we have $\left|\dfrac{-11}{5}\right| \times \sqrt{|-25|} = \dfrac{11}{5} \times \sqrt{25} = \dfrac{11}{5} \times 5 = 11$.

Example 6

Find the absolute value for each of the following.

(a) zero

Since 0 is neither positive or negative, $|0| = 0$.

(b) 4

4 is already positive, so $|4| = 4$.

(c) $-\pi$

π is often used to represent a positive constant useful in making calculations about circles; however, its precise value is irrelevant here, since we are seeling only to find the absolute value of its opposite. Since $-\pi$ is negative, $|-\pi| = \pi$.

(d) a, where a is a real number

If a represents a real number, then it

could be positive, negative, or zero. Therefore, to properly find the absolute value for a, we need to consider all possibilities. So, for $a > 0$, $|a| = a$; for $a = 0$; $|a| = 0$; for $a < 0$, $|a| = -a$.

$$\text{i.e., } |a| = \begin{cases} a \text{ if } a > 0 \\ 0 \text{ if } a = 0 \\ a \text{ if } a < 0 \end{cases}$$

Questions

Identify the parts of the following expressions.

1. $6x^2y$

2. $4 + \dfrac{3y}{z}$

3. $\dfrac{ab}{yz^4}$

Specify whether each of the following expressions is a monomial, binomial, or trinomial.

4. $6x^3 - \dfrac{3y}{4}$

5. $4y^4z$

6. $15 - x^3y + y^2$

Simplify the following.

7. $4 + (-7) + 2 + (-5)$.

8. $144 + (-317) + 213$.

9. $|43 - 62| - |-17 - 3|$.

10. $91,203 - 37,904 + 1,073$.

11. $|-42| \times |7|$.

12. $-6 \times 5 \times -10 \times -4 \times 0 \times 2$.

13. $(-180) \div (-12)$.

14. $|-76| \div |-4|$.

15. $\dfrac{4 + 8 \times 2}{5 - 1}$

Solve the following.

16. $96 \div 3 \div 4 \div 2 = $ _____

17. $3 + 4 \times 2 - 6 \div 3 = $ _____

18. $[(4 + 8) \times 3] \div 9 = $ _____

19. $\dfrac{11 \times 2 + 2}{16 - 2 \times 2} = $ _____

20. $|-8 - 4| \div 3 \times 6 + (-4) = $ _____

Answers

1. In $6x^2y$, there is one term which has a coefficient, 6, a variable, x, and another variable, y.

2. In $4 + \dfrac{3y}{z}$, there are two terms. The first term has only a coefficient, 4, and the second term has a coefficient, 3, and two variables, y and z.

3. In $\dfrac{ab}{yz^4}$, there is one term that has four variables, a, b, y, and z.

4. This expression has two terms, $6x^3$ and $3y/4$. Therefore, this expression is a binomial.

5. Although this expression has several parts, it has only one term, $4y^4z$. Therefore, this expression is a monomial.

6. This expression has three terms, 15, x^3y, and y^2. Therefore, this expression is a trinomial.

7. Adding -7 is the same as subtracting positive 7, so first subtract $4 - 7 = -3$, then add $-3 + 2 = -1$, and finally $-1 + (-5) = -6$.

8. Adding -317 is the same as subtracting positive 317, so first subtract $144 - 317 = -173$. Then add $-173 + 213 = 213 - 173 = 40$.

9. Subtract $|43 - 62|$ and $|-17 - 3|$ first, as

you would if parentheses were used instead of the vertical absolute value lines, so that we get | –19 | – | –20 |. Next solve | –19 | and | –20 |, to give us 19 – 20 = –5.

10. Subtract 91,203 – 37,904 first, to get 53,299 + 1,073 = 54,372.

11. First take the absolute values of each term, leaving 42 × 7 = 294.

12. Rather than multiplying these terms, we can save time by recalling that multiplying zero and any number gives us zero. For clarity, we may regroup the equation as follows:

$$(-6 \times 5 \times -10 \times -4 \times 2) \times 0 = 0$$

13. Recall that to divide two numbers having like signs, we divide their absolute values and write the result with a positive sign.

$$(-180) \div (-12) = |-180| \div |-12|$$
$$= 180 \div 12 = 15$$

14. | –76 | ÷ | –4 | = 76 ÷ 4 = 19

15. First multiply 8 × 2 = 16, then sum the numerator and subtract in the denominator, to get 4 + 16 = 20 and 5 – 1 = 4. Then divide 20 ÷ 4 = 5.

16. Perform the division in order from left to right. Begin by dividing 96 ÷ 3 = 32, then 32 ÷ 4 = 8, and finally 8 ÷ 2 = 4.

17. First multiply 4 × 2 = 8 and divide 6 ÷ 3 = 2. Then sum 3 + 8 – 2 = 9.

18. First perform the operations within the innermost parenthesis, so (4 + 8) = 12. Then perform the operations within the brackets, [(12) × 3] = 36. Finally, divide [36] ÷ 9 = 4.

19. First multiply 11 × 2 in the numerator and 2 × 2 = 4 in the denominator. Next add 22 + 2 = 24 in the numerator, and subtract 16 – 4 = 12 in the denominator to get the fraction 24/12 = 2.

20. First subtract | –8 – 4 |, since vertical absolute value lines are used like parenthesis, so

that we get | –12 | ÷ 3 × 6 + (–4). Then find the absolute value of | –12 | = 12, giving us 12 ÷ 3 × 6 + (–4). Next, divide 12 ÷ 3 = 4, then multiply 4 × 6 = 24, and then add 24 + (–4) = 24 – 4 = 20.

MULTIPLICATION AND DIVISION OF MONOMIALS

The following demonstrates how to perform some mathematical operations you already know on monomials involving exponents. Note that when two terms have the same variable, they are said to be "like terms."

Monomial Operations

1. When multiplying like terms, $a^p \times a^q = a^p a^q = a^{p+q}$.

2. When dividing like terms, $a^p/a^q = a^{p-q}$.

3. For exponents outside parenthesis in monomial multiplication, $(a \times b)^p = (ab)^p = a^p b^p$.

4. For exponents outside parenthesis in monomial division, $(a/b)^p = a^p/b^p$, $b \neq 0$.

5. For exponents outside parenthesis surrounding a monomial with an exponent, $(a^p)^q = a^{pq}$.

Example 1

Simplify the following expressions.

(a) $x^2 x^3$

The variable does not change when multiplying like terms, but the exponents are added together. Here, x is unchanged while we add 2 + 3 = 5. So $x^2 x^3 = x^5$.

(b) $2y^6 y^2$

The variable does not change when multiplying like terms, but the exponents are added together. Here, 2 is the only term of its kind and y is unchanged while we add $6 + 2 = 8$. So $2y^6y^2 = 2y^8$.

(c) x^3x

The variable does not change when multiplying like terms, but the exponents are added together. Since the second term, x, has no exponent shown, we assume the exponent is 1. Here, x is unchanged while we add $3 + 1 = 4$. So $x^3x = x^4$.

(d) $x^2y^4x^3$

The variable does not change when multiplying like terms, but the exponents are added together. Remember that if the terms have different variables, they remain separate. For like terms, the exponents are still added together. Here, x is unchanged while we add $2 + 3 = 5$, and y^4 is unchanged. So $x^2y^4x^3 = x^5y^4$.

Example 2

Divide the following monomials.

(a) y^3/y^2

Just as the variable does not change when multiplying like terms, it remains the same while dividing like terms. The exponents are subtracted. Here, y is unchanged while we subtract $3 - 2 = 1$. So $y^3/y^2 = y^1 = y$.

(b) $z^5/4z^3$

Just as the variable does not change when multiplying like terms, it remains the same while dividing like terms. The exponents are subtracted. Here, y is unchanged while we subtract $5 - 3 = 2$. Also notice the 4 in the denominator; since there is no integer

in the numerator, we assume a one. We can think of this problem as $1/4 \times z^5/z^3 = 1/4 \, z^2$ or $z^2/4$.

(c) $6x^5y/3y^2z^2$

Just as the variable does not change when multiplying like terms, it remains the same while dividing like terms. The exponents are subtracted. Here, y is unchanged while we subtract $1 - 2 = -1$. Remember that if the terms have different variables, they remain separate. There is only one x term and only one z term, so they remain unchanged. Also notice the 6 in the numerator and the 3 in the denominator, a fraction which we can simplify as follows: $6/3 = 2$. We can think of this problem in parts, as $2 \times x^5/z^2 \times y/y^2 = 2x^5y^{-1}/z^2$.

Example 3

Simplify the following expressions.

(a) $(xy)^4$

Recall that when a term is shown with an exponent, the exponent indicates the number of times that the term should be multiplied by itself. In the case of a monomial within parenthesis that have an exponent, we can rewrite the monomial as multiplication problem: $(xy)^4 = (xy \times xy \times xy \times xy) = (x \times x \times x \times x) \times (y \times y \times y \times y) = x^4y^4$.

(b) $(3y)^3$

Recall that when a term is shown with an exponent, the exponent indicates the number of times that the term should be multiplied by itself. In the case of a monomial within parenthesis that have an exponent, we can rewrite the monomial as multiplication problem: $(3y)^3 = (3y \times 3y \times 3y) = (3 \times$

$3 \times 3) \times (y \times y \times y) = 27y^3$.

OPERATIONS WITH POLYNOMIALS

The four basic operations of polynomials are addition, subtraction, multiplication, and division. For each of these operations, the results may be verified by replacing any variable with a constant. Remember that $x = 1x$, not necessarily 1.

Example 1

Simplify the following polynomials.

(a) $(3x^2 - 7) + (x^2 + 2)$

When adding polynomials, remember to sum the coefficients of like terms. In this case, we have $3x^2 + 1x^2 = 4x^2$ and $(-7) + 2 = -5$. So the answer is $4x^2 - 5$.

(b) $(x^2 - 5x + 1) + (2x^2 - 3)$

When adding polynomials, remember to sum the coefficients of like terms. In this case, we have $1x^2 + 2x^2 = 3x^2$ and $1 + (-3) = -2$. $-5x$ remains the same, so the answer is $3x^2 - 5x - 2$.

(c) $(2x^2 - 3) - (11x^2 - 3)$

Subtraction of polynomials is similar to addition, but we must first distribute the negative sign to the entire polynomial to which it is applied. Change the signs of all terms in the expression that is being subtracted, and then add this result to the other expression. So, $(2x^2 - 3) - (11x^2 - 3)$ becomes $(2x^2 - 3) + (-11x^2 + 3)$. Adding them together, we have $2x^2 - 11x^2 = -9x^2$ and $(-3) + 3 = 0$. So the answer is $-9x^2 + 0 = -9x^2$.

(d) $(4x^2 + 3x - 8) - (3x^2 - 7x + 4)$

Subtraction of polynomials is similar to addition, but we must first distribute the negative sign to the entire polynomial to which it is applied. Change the signs of all terms in the expression that is being subtracted, and then add this result to the other expression. So, $(4x^2 + 3x - 8) - (3x^2 - 7x + 4)$ becomes $(4x^2 + 3x - 8) + (-3x^2 + 7x - 4)$. Adding them together, we have $4x^2 - 3x^2 = 1x^2 = x^2$, $3x + 7x = 10x$, and $(-8) - 4 = -12$. So the answer is $x^2 + 10x - 12$.

Example 2

Multiply the following expressions.

(a) $(2xy)(-5x^2y^3)$

In multiplication of monomials, multiply the coefficients, but add the exponents of like variables. So multiply $2 \times -5 = -10$, $x \times x^2 = x^3$, and $y \times y^3 = y^4$. The answer is $-10x^3y^4$.

(b) $(7l^2m)(6k^2l^2)$

In multiplication of monomials, multiply the coefficients, but add the exponents of like variables. So multiply $7 \times 6 = 42$ and $l^2 \times l^2 = l^4$. Therefore, the answer is $42k^2l^4m$.

(c) $(9q)(3q^2 - 4qr)$

In multiplication of a monomial and a polynomial, multiply the monomial by each term of the polynomial. Remember, as with multiplication of monomials, to multiply coefficients, but add the exponents of like variables. It may be easy to break the problem into parts. First find $9x \times 3q^2$ and then add the result to the product of $9q \times -4xy$. So multiply $9 \times 3 = 27$ and $q \times q^2 = q^3$; then multiply $9 \times -4 = -36$ and $q \times q = q^2$. Therefore, the answer is $27q^3 +$

$(-36q^2r) = 27q^3 - 36q^2r$.

(d) $(6x - 1)(2x + 4)$

In multiplication of binomials, multiply each term by the term in the other parentheses one time each. An easy way to do this is to remember an acronym called FOIL. FOIL stands for "first, outer, inner, last," which is the order we can multiply our terms. Multiply the "first" terms – in this case, $6x \times 2x = 12x^2$; then the "outer" terms, which is to say the first term from the first binomial and the last term from the second binomial, giving us $6x \times 4 = 24x$; then the "inner" terms, $-1 \times 2x = -2x$; and then the "last" terms, $-1 \times 4 = -4$. The last step is to add our terms: $12x^2 + 24x + (-2x) + -4 = 12x^2 + 22x - 4$.

(e) $(4ps - 5p)(2p + 4s + 4)$

In multiplication of polynomials, multiply each term by the term in the other parentheses one time each. An easy way to do this is to multiply the first term of the smallest polynomial by the terms of the other polynomial first, and then move on to the others. In this case, we would begin by multiplying $4ps \times 2p = 8p^2s$, then $4ps \times 4s = 16ps^2$, then $4ps \times 4 = 16ps$; and then we would move on to $-5p \times 2p = -10p^2$, then $-5p \times 4s = -20ps$, then $-5p \times 4 = -20p$. The next step is to add our terms: $8p^2s + 16ps^2 + 16ps + (-10p^2) + (-20ps) + (-20p)$. Since we can add terms with like variables, group all of the like variables together, giving us $8p^2s + 16ps^2 + 16ps + (-20ps) + (-10p^2) + (-20p) = 8p^2s + 16ps^2 + (16ps - 20ps) - 10p^2 - 20p = 8p^2s + 16ps^2 - 4ps - 10p^2 - 20p$.

Example 3

Divide the following expressions.

(a) $20v^3w^5 \div 4vw^2$

In division of monomials, divide the coefficients, but subtract the exponents of like variables. So divide $20 \div 4 = 5$, $v^2 \div v = v$, and $w^5 \div w^2 = w^3$. The answer is $5vw^3$.

(b) $8r^4s^2t^5 \div (4rst)^2$

Remember order of operations. First, $(4rst)^2 = 16r^2s^2t^2$. Next divide, recalling that in division of monomials, divide the coefficients, but subtract the exponents of like variables. So divide $8 \div 16 = \frac{1}{2}$, $r^4 \div r^2 = r^2$, $s^2 \div s^2 = 1$, and $t^5 \div t^2 = t^3$. The answer is $8r^2t^3$.

(c) $(2x^2 + 6 + x) \div (1 + x)$

$$x + 1 \overline{)2x^2 + x + 6} \quad \longrightarrow$$

$$\begin{array}{r} 2x \quad\quad\quad\quad \\ x + 1 \overline{)\ 2x^2 + x + 6} \\ -2x^2 + 2x \quad \longrightarrow \\ \hline -x \end{array}$$

$$\begin{array}{r} 2x - 1 + \dfrac{7}{x+1} \\ x + 1 \overline{)\ 2x^2 + x + 6} \\ -2x^2 + 2x \\ \hline -x + 6 \\ -(-x - 1) \\ \hline 7 \end{array}$$

In division of a polynomial by a polynomial, we must follow the procedure of long division. First, arrange the terms of both polynomials in order of ascending or descending powers of one variable. Here, we'll arrange in terms of descending powers of x, so that we have $(2x^2 + x + 6) \div (x + 1)$. The first term of the dividend is divided by the first term of the divisor, which gives the first term of the quotient, $2x$. Then, as in long division of whole numbers and decimals, the first term of the quotient is multiplied by the entire divisor and the result, $2x^2 + 2x$, is subtracted from the dividend. Using the remainder from this subtraction, repeat those three steps (divide the dividend by the next term of the divisor, multiply the resulting quotient term by the entire divisor, and subtract the result from the dividend) until the degree of the remainder is less than the degree of the divisor. The result is written as follows:

$$\frac{\text{dividend}}{\text{divisor}} = \text{quotient} + \frac{\text{remainder}}{\text{divisor}}$$

divisor $\neq 0$

So, $(2x^2 + x + 6) \div (x + 1) = 2x - 1 + \dfrac{7}{x+1}$.

(d) $(4x + 3x^2 + 7 + 2x) \div (x^2 + 3x)$

$$x^2 + 3x \,\overline{\smash{\big)}\, 3x^3 + 2x^2 + 4x + 7} \longrightarrow$$

$$\begin{array}{r} 3x \\ x^2 + 3x \,\overline{\smash{\big)}\, 3x^3 + 2x^2 + 4x + 7} \\ \underline{-(3x^3 + 9x^2)} \\ -7x \end{array} \longrightarrow$$

$$\begin{array}{r} 3x - 7 \\ x^2 + 3x \,\overline{\smash{\big)}\, 3x^3 + 2x^2 + 4x + 7} \\ \underline{-(3x^3 + 9x^2)} \\ -7x + 4x \\ \underline{-(-7x^2 - 21x)} \\ 25x \end{array} \longrightarrow$$

$$\begin{array}{r} 3x - 7 + \dfrac{25x + 7}{x^2 + 3x} \\ x^2 + 3x \,\overline{\smash{\big)}\, 3x^3 + 2x^2 + 4x + 7} \\ \underline{-(3x^3 + 9x^2)} \\ -7x + 4x \\ \underline{-(-7x^2 - 21x)} \\ 25x + 7 \end{array} \longrightarrow$$

In division of a polynomial by a polynomial, we must follow the procedure of long division. First, arrange the terms of both polynomials in order of ascending or descending powers of one variable. Here, we'll arrange in terms of descending powers of x, so that we have $(3x^3 + 2x^2 + 4x + 7) \div (x^2 + 3x)$. The first term of the dividend is divided by the first term of the divisor, which gives the first term of the quotient, $3x$. Then, as in long division of whole numbers and decimals, the first term of the quotient is multiplied by the entire divisor and the result, $3x^3 + 9x$, is subtracted from the dividend. Using the remainder from this subtraction, $-7x^2$, repeat those three steps (divide the dividend by the next term of the divisor, multiply the resulting quotient term by the entire divisor, and subtract the result from the dividend) until the degree of the remainder is less than the degree of the divisor. So,

$$(3x^3 + 2x^2 + 4x + 7) \div (x^2 + 3x)$$

$$= 2x - 1 + \frac{25x + 7}{x^2 + 3x}.$$

Questions

Simplify the following polynomials.

1. $(7a^2 - 3c) + (8a^2 + c)$

2. $(4x + 3y - 6w) - (8w + 5y - 3x)$

3. $-(5x^2y + 2xy^2) + (10xy - 8x^2y + 6xy^2)$

Multiply the following monomials and polynomials.

4. $(8r^2t^3)(7r^4t)$

5. $(-6m^5n^4)(-5m^5n)$

6. $(-3t)(t^2 + 4t - 2)$

7. $(2n - x)(5n + 3x)$

8. $(x^2 - 2xy - y^2)(x + 4y)$

9. $(2x^2 + 11x - 2)(x + 9)$

Divide the following monomials and polynomials.

10. $28a^8c^4 \div 4a^2c^3$

11. $-90x^{12}y^7 \div 15x^4y^5$

12. $(5y^2 + 3y - 8) \div (y - 1)$

13. $(6p) + 9pn + 3n^2) \div (p + n)$

14. $(m3 + m^2 - 5m - 3) \div (m + 2)$

15. $(4m^2 + 7m + 12) \div (m - 1)$

Answers

1. $15a^2 - 2c$

$$7a^2 - 3c + 8a^2 + c$$

$$= 7a^2 + 8a^2 - 3c + c$$

$$= 15a^2 - 2c$$

2. $7x - 2y - 14w$

$$4x + 3y - 6w - 8w - 5y + 3x$$

$$= 4x + 3x + 3y - 5y - 6w - 8w$$

$$= 7x - 2y - 14w$$

3. $-13x^2y + 4xy^2 + 10xy$

$$-5x^2y - 2xy^2 + 10xy - 8x^2y + 6xy^2$$

$$= -5x^2y - 8x^2y - 2xy^2 + 6xy^2 + 10xy$$

$$= -13x^2y + 4xy^2 + 10xy$$

4. $56r^6t^4$

$$(8r^2t^3)(7r^4t)$$

$$= (8 \times 7)(r^2 \times r^4)(t^3 \times t)$$

$$= 56r^6t^4$$

5. $30m^{10}n^5$

$$(-6m^5n^4)(-5m^5n)$$

$$= (-6 \times -5)(m^5 \times m^5)(n^4 \times n)$$

$$= 30m^{10}n^5$$

6. $-3t^3 - 12t^2 + 6t$

$$(-3t)(t^2 + 4t - 2)$$

$$= (-3t)(t^2) + (-3t)(4t) + (-3t)(-2)$$

$$= -3t^3 - 12t^2 + 6t$$

7. $10n^2 + nx - 3x^2$

$$(2n - x)(5n + 3x)$$

$$= (2n)(5n) + (2n)(3x) + (-x)(5n) + (-x)(3x)$$

$= 10n^2 + nx - 3x^2$

8. $x^3 + 2x^2y - 9xy^2 - 4y^3$

$(x^2 - 2xy - y^2)(x + 4y)$

$= (x^2)(x) + (x^2)(4y) + (-2xy)(x) +$
$(-2xy)(4y) + (-y^2)(x) +$
$(-y^2)(4y)$

$= x^3 + (4x^2y - 2x^2y) + (-8xy^2 - xy^2) -$
$4y^3$

$= x^3 + 2x^2y - 9xy^2 - 4y^3$

9. $2x^3 + 29x^2 + 97x - 18$

$(2x^2 + 11x - 2)(x + 9)$

$= (2x^2)(x) + (2x^2)(9) + (11x)(x) +$
$(11x)(9) + (-2)(x) + (-2)(9)$

$= 2x^3 + 18x^2 + (99x - 2x) - 18$

$= 2x^3 + 29x^2 + 97x - 18$

10. $7a^6c$

$28a^8c^4 \div 4a^2c^3$

$= (28 \div 4)(a^8 \div a^2)(c^4 \div c^3)$

$= 7a^6c$

11. $-6x^8y^2$

$-90x^{12}y^7 \div 15x^4y^5$

$= (-90 \div 15)(x^{12} \div x^4)(y^7 \div y^5)$

$= -6x^8y^2$

12. $5y + 8$

$(5y^2 + 3y - 8) \div (y - 1)$

$$
\begin{array}{r}
5y + 8 \\
y - 1 \overline{\smash{)}\, 5y^2 + 3y - 8} \\
-(5y^2 - 5y) \\
\hline
8y - 8 \\
-(8y - 8) \\
\hline
0
\end{array}
$$

13. $6p + 3n$

$(6p^2 + 9pn + 3n^2) \div (p + n)$

$$
\begin{array}{r}
6p + 3n \\
p + n \overline{\smash{)}\, 6p^2 + 9pn + 3n^2} \\
-(6p^2 + 6pn) \\
\hline
3pn + 3n^2 \\
-(3pn + 3n^2) \\
\hline
0
\end{array}
$$

14. $m^2 - m - 3 + \dfrac{3}{m+2}$

$(m^3 + m^2 - 5m - 3) \div (m + 2)$

$$
\begin{array}{r}
m^2 - m - 3 + \dfrac{3}{m+2} \\
m + 2 \overline{\smash{)}\, m^3 + m^2 - 5m - 3} \\
-(m^3 + 2m^2) \\
\hline
-m^2 - 5m \\
-(-m^2 - 2m) \\
\hline
-3m - 3 \\
-(-3m - 6) \\
\hline
3
\end{array}
$$

15. $4m + 11 + \dfrac{23}{m-1}$

$(4m^2 + 7m + 12) \div (m - 1)$

$$
\begin{array}{r}
4m + 11 + \dfrac{23}{m-1} \\
m - 1 \overline{\smash{)}\, 4m^2 + 7m + 12} \\
-(4m^2 - 4m) \\
\hline
11m + 12 \\
-(11m - 11) \\
\hline
23
\end{array}
$$

SIMPLIFYING ALGEBRAIC EXPRESSIONS

To factor a polynomial completely is to find the prime factors of the polynomial with respect to a specified set of numbers.

The following concepts are important while factoring or simplifying expressions.

1. The factors of an algebraic expression consist of two or more algebraic expressions which, when multiplied together, produce the given algebraic expression.

2. A **prime factor** is a polynomial with no factors other than itself and 1. The **least common multiple (LCM)** for a set of numbers is the smallest quantity divisible by every number of the set. For algebraic expressions the least common numerical coefficients for each of the given expressions will be a factor.

3. The **greatest common factor (GCF)** for a set of numbers is the largest factor that is common to all members of the set.

4. For algebraic expressions, the greatest common factor is the polynomial of highest degree and the largest numerical coefficient which is a factor of all the given expressions.

Some important formulas, useful for the factoring of polynomials are listed below.

$$a(c + d) = ac + ad$$

$$(a + b)(a - b) = a^2 - b^2$$

$$(a + b)(a + b) = (a + b)^2 = a^2 + 2ab + b^2$$

$$(a - b)(a - b) = (a - b)^2 = a^2 - 2ab + b^2$$

$$(x + a)(x + b) = x^2 + (a + b)x + ab$$

$$(ax + b)(cx + d) = acx^2 + (ad + bc)x + bd$$

$$(a + b)(c + d) = ac + bc + ad + bd$$

$$(a + b)(a + b)(a + b) = (a + b)^3$$

$$= a^3 + 3a^2b + 3ab^2 + b^3$$

$$(a - b)(a - b)(a - b) = (a - b)^3$$
$$= a^3 - 3a^2b + 3ab^2 - b^3$$

$$(a - b)(a^2 + ab + b^2) = a^3 - b^3$$

$$(a + b)(a^2 - ab + b^2) = a^3 + b^3$$

$$(a + b + c)^2 = a^2 + b^2 + c^2 + 2ab + 2ac + 2bc$$

$$(a - b)(a^3 + a^2b + ab^2 + b^3) = a^4 - b^4$$

$$(a - b)(a^4 + a^3b + a^2b^2 + ab^3 + b^4) = a^5 - b^5$$

$$(a - b)(a^5 + a^4b + a^3b^2 + a^2b^3 + ab^4 + b^5) = a^6 - b^6$$

$$(a - b)(a^{n-1} + a^{n-2}b + a^{n-3}b^2 + \ldots + ab^{n-2} + b^{n-1}) = a^n - b^n$$

where n is any positive integer (1, 2, 3, 4, ...).

$$(a + b)(a^{n-1} - a^{n-2}b + a^{n-3}b^2 - \ldots - ab^{n-2} + b^{n-1}) = a^n + b^n$$

where n is any positive odd integer (1, 3, 5, 7, ...).

The procedure for factoring an algebraic expression completely is as follows:

Step 1: First find the greatest common factor if there is any. Then examine each factor remaining for greatest common factors.

Step 2: Continue factoring the factors obtained in Step 1 until all factors other than monomial factors are prime.

Example 1

Factor the following.

(a) $2xy$.

$2xy$ has the factors 2, x, y, $2x$, xy, and $2y$, in addition to itself and 1.

(b) x.

x is considered a prime, since it can only be divided by itself and 1.

(c) $x^3y - x^2y^2 + xy^4$

$x^3y - x^2y^2 + xy^4 = (xy)(x^2 - xy + y^3)$ by using the greatest common term, then dividing it into all 3 original terms.

(d) $27 - n^3$.

Rewrite as $(3)^3 - (n)^3$ and use the formula for factoring the difference of 2 cubes.

Then $(3)^3 - (n)^3 = (3 - n)(9 + 3n + n^2)$

(e) $2x^2 + 11x + 5$

This is an example of "trial and error" trinomial factoring into 2 binomials.

$2x^2 + 11x + 5 = (2x + 1)(x + 5)$

(f) $4 - 16x^2$.

$4 - 16x^2 = 4(1 - 4x^2)$
$= 4(1 + 2x)(1 - 2x)$

Example 2

Find the LCM of 4 and 6.

The LCM of 4 and 6 is 12, since 12 is divisible by both 4 and 6, and it is the smallest number with that property.

Example 3

(a) Find the GCF of 9 and 12.

The GCF of 9 and 12 is 3, since 3 divides evenly into each of 9 and 12, and no larger number has this property.

(b) Find the GCF of $4x^3$, $10x^4$, and $14x^5$.

If a polynomial is given by $4x^3 + 10x^4 + 14x^5$, then $2x^3$ is the GCF of each term.

Example 4

Express each of the following in the fewest terms possible.

(a) $3x^2 + 2x^2 - 4x^2$

Factor x^2 in the expression.

$3x^2 + 2x^2 - 4x^2 = (3 + 2 - 4)x^2$
$= 1x^2 = x^2$.

(b) $5axy^2 - 7axy^2 - 3xy^2$

Factor xy^2 in the expression and then factor a.

$5axy^2 - 7axy^2 - 3xy^2 = (5a - 7a - 3)xy^2$

$= [(5 - 7)a - 3]xy^2$

$= (-2a - 3)xy^2$.

Questions

Factor each expression

1. $6t^5 + 10t^3$

2. $49m^2n^2 - 16$

3. $y^2 - 11y + 28$

4. $x^2 + 10xy - 24y^2$

5. $9z^2 + 6z + 1$

6. $r^3 + 8$

7. $a^4 - 1$

8. $6c^2 + 11c - 10$

9. $12x^3 - 27x$

10. $p^2t^2 + 13pt + 40$

Answers

1. $(2t^3)(3t^2 + 5)$

$6t^5 + 10t^3 = (2t^3)(3t^2 + 5)$

2. $(7mn + 4)(7mn - 4)$

$49m^2n^2 - 16 = (7mn + 4)(7mn - 4)$

3. $(y - 4)(y - 7)$

$y^2 - 11y + 28 = (y - 4)(y - 7)$

4. $(x - 2y)(x + 12y)$

$x^2 + 10xy - 24y^2 = (x - 2y)(x + 12y)$

5. $(32 + 1)(3z + 1)$ or $(32 + 1)^2$

$9z^2 + 6z + 1 = (3z + 1)(3z + 1)$

6. $(r + 2)(r^2 - 2r + 4)$

$r^3 + 8 = (r + 2)(r^2 - 2r + 4)$

7. $(a^2 + 1)(a - 1)(a + 1)$

$a^4 - 1 = (a^2 + 1)(a - 1)(a + 1)$

8. $(3c - 2)(2c + 5)$

$6c^2 + 11c - 10 = (3c - 2)(2c + 5)$

9. $(3x)(2x - 3)(2x + 3)$

$12x^3 - 27x = (3x)(2x - 3)(2x + 3)$

10. $(pt + 5)(pt + 8)$

$p^2t^2 + 13pt + 40 = (pt + 5)(pt + 8)$

EQUATIONS

An **equation** is defined as a statement in which two separate expressions are equal.

A **solution** to an equation containing a single variable is a number that makes the equation true when it is substituted for the variable. For example, in the equation $3x = 18$, 6 is the solution since $3(6) = 18$. Depending on the equation, there can be more than one solution. Equations with the same solutions are said to be **equivalent equations**. An equation without a solution is said to have a solution set that is the **empty** or **null** set and is represented by ϕ.

Replacing an expression within an equation by an equivalent expression will result in a new equation with solutions equivalent to the original equation. Suppose we are given the equation

$3x + y + x + 2y = 15$.

By combining like terms we get

$3x + y + x + 2y = 4x + 3y$.

Since these two expressions are equivalent, we can substitute the simpler form into the equation to get

$4x + 3y = 15$

Performing the same operation to both sides of an equation by the same expression will result in a new equation that is equivalent to the original equation.

1.   Addition or subtraction

$y + 6 = 10$

we can add (-6) to both sides

$y + 6 + (-6) = 10 + (-6)$

to get $y + 0 = 10 - 6 \rightarrow y = 4$

2.   Multiplication or division

$3x = 6$

$3x \div 3 = 6 \div 3$

$x = 2$

$3x = 6$ is equivalent to $x = 2$.

3.   Raising to a power

$\sqrt{x + 2} = 9$

$\left(\sqrt{x + 2}\right)^2 = 9^2$

$x + 2 = 81$

$x = 79$

LINEAR EQUATIONS

A linear equation with one unknown is one that can be put into the form $ax + b = 0$, where a and b are constants, $a \neq 0$.

To solve a linear equation means to transform it in the form $x = {}^{-b}/_a$.

If the equation has unknowns on both sides of the equality, it is convenient to put similar terms on the same sides, e.g.,

$$4x + 3 = 2x + 9$$
$$4x + 3 - 2x = 2x + 9 - 2x$$
$$(4x - 2x) + 3 = (2x - 2x) + 9$$
$$2x + 3 = 0 + 9$$
$$2x + 3 - 3 = 0 + 9 - 3$$
$$2x = 6$$
$${}^{2x}/_2 = {}^6/_2$$
$$x = 3.$$

If the equation appears in fractional form, it is necessary to transform it, using cross multiplication. Cross-multiplication simply means that for an equation where each side is a fraction, multiply both sides by both denominators. In other words, for any equation $a/b = c/d$ (b, $d \neq 0$), we get $ad = bc$. So, repeating the same procedure as above, we obtain:

$$\frac{3x + 4}{3} \bowtie \frac{7x + 2}{5}$$

By using cross-multiplication we would obtain:

$$3(7x + 2) = 5(3x + 4).$$

This is equivalent to:

$$21x + 6 = 15x + 20,$$

which can be solved as above:

$$21x + 6 = 15x + 20$$
$$21x - 15x + 6 = 15x - 15x + 20$$

$$6x + 6 - 6 = 20 - 6$$
$$6x = 14$$
$$x = {}^{14}/_6$$
$$x = {}^7/_3$$

If there are radicals in the equation, it is necessary to square both sides and then proceed as above:

$$\sqrt{3x + 1} = 5$$
$$\left(\sqrt{3x + 1}\right)^2 = 5^2$$
$$3x + 1 = 25$$
$$3x + 1 - 1 = 25 - 1$$
$$3x = 24$$
$$x = {}^{24}/_3$$
$$x = 8$$

Example 1

Solve the equation
$$2(x + 3) = (3x + 5) - (x - 5).$$

We transform the given equation to an equivalent equation in which we can easily recognize the solution set.

$2(x + 3) = 3x + 5 - (x - 5)$

Distribute, $2x + 6 = 3x + 5 - x + 5$

Combine terms, $2x + 6 = 2x + 10$

Subtract $2x$ from both sides, $6 = 10$

Since $6 = 10$ is not a true statement, there is no real number x which will make the original equation true. The equation is inconsistent and the solution set is ϕ, the empty set.

Example 2

Solve the equation
$$2({}^2/_3\, y + 5) + 2(y + 5) = 130.$$

The procedure for solving this equation is as follows:

$^4/_3y + 10 + 2y + 10 = 130$,
Distributive property

$^4/_3y + 2y + 20 = 130$,
Combining like terms

$^4/_3y + 2y = 110$,
Subtracting 20 from both sides

$^4/_3y + ^6/_3y = 110$,
Converting $2y$ into a fraction with denominator 3

$^{10}/_3y = 110$,
Combining like terms

$y = 110 \times {}^3/_{10} = 33$,

Dividing by $^{10}/_3$

Check: Replace y with 33 in the original equation,

$2(^2/_3(33) + 5) + 2(33 + 5) = 130$

$2(22 + 5) + 2(38) = 130$

$2(27) + 76 = 130$

$54 + 76 = 130$

$130 = 130$

Therefore the solution to the given equation is $y = 33$.

Questions

Solve for x.

1. $4x - 2 = 10$

2. $7x + 1 - x = 2x - 7$

3. $\dfrac{1}{3}x + 3 = \dfrac{1}{2}x$

4. $0.4x + 1 = 0.7x - 2$

5. $4(3x + 2) - 11 = 3(3x - 2)$

6. $6x + 9 = -15$

7. $\dfrac{1}{3}x + 1 = \dfrac{1}{4}x + 5$

8. $0.09x - 4 = 0.04x$

9. $7(x + 2) = 3(3x + 5)$

10. $\sqrt{x + 3} = 6$

Answers

1. $x = 3$
$$4x - 2 + 2 = 10 + 2$$
$$4x \div 4 = 12 \div 4$$
$$x = 3$$

2. $x = -2$
$$7x + 1 - x = 2x - 7$$
$$6x + 1 - 2x - 1 = 2x - 7 - 2x - 1$$
$$4x \div 4 = -8 \div 4$$
$$x = -2$$

3. $x = 18$
$$\frac{1}{3}x + 3 - \frac{1}{3}x = \frac{1}{2}x - \frac{1}{3}x$$
$$3 = \frac{3}{6}x - \frac{2}{6}x$$
$$3 \times 6 = \frac{1}{6}x \times 6$$
$$18 = x$$

4. $x = 10$
$$0.4x + 1 - 0.4x + 2 = 0.7x - 2 - 0.4x + 2$$
$$3x \div 0.3 = 0.3x \div 0.3$$
$$x = 10$$

5. $x = -1$
$$4(3x + 2) - 11 = 3(3x - 2)$$
$$12x + 8 - 11 = 9x - 6$$
$$12x - 3 - 9x + 3 = 9x - 6 - 9x + 3$$

$$3x \div 3 = -3 \div 3$$
$$x = -1$$

6. $x = -4$

$$6x + 9 - 9 = -15 - 9$$
$$6x \div 6 = -24 \div 6$$
$$x = -4$$

7. $x = 48$

$$\frac{1}{3}x + 1 - \frac{1}{4}x - 1 = \frac{1}{4}x + 5 - \frac{1}{4}x - 1$$
$$\frac{1}{3}x - \frac{1}{4}x = 4$$
$$\frac{1}{12}x = 4$$
$$x = 48$$

8. $x = 80$

$$0.09x - 4 - 0.09x = 0.04x - 0.09x$$
$$-4 \div 0.05 = -0.05x \div 0.05$$
$$80 = x$$

9. $x = -\frac{1}{2}$

$$7(x + 2) = 3(3x + 5)$$
$$7x + 14 - 7x - 15 = 9x + 15 - 7x - 15$$
$$-1 \div 2 = 2x \div 2$$
$$-\frac{1}{2} = x$$

10. $x = 33$

$$\sqrt{x + 3} = 6$$
$$(\sqrt{x + 3})^2 = 6^2$$
$$x + 3 - 3 = 36 - 3$$
$$x = 33$$

INEQUALITIES

An inequality is a statement where the value of one quantity or expression is greater than (>), less than (<), greater than or equal to (≥), less than or equal to (≤), or not equal to (≠) that of another.

For example, the expression 5 > 4 means that the value of 5 is greater than the value of 4.

A **conditional inequality** is an inequality whose validity depends on the values of the variables in the sentence. That is, certain values of the variables will make the sentence true, and others will make it false. $3 - y > 3 + y$ is a conditional inequality for the set of real numbers, since it is true for any replacement less than zero and false for all others.

$x + 5 > x + 2$ is an **absolute inequality** for the set of real numbers, meaning that for any real value x, the expression on the left is greater than the expression on the right.

$5y < 2y + y$ is inconsistent for the set of non-negative real numbers. For any y greater than or equal to 0, the sentence is always false. A sentence is inconsistent if it is always false when its variables assume allowable values.

The solution of a given inequality in one variable x consists of all values of x for which the inequality is true.

The graph of an inequality in one variable is represented by either a ray or a line segment on the real number line.

The endpoint is not a solution if the variable is strictly less than or greater than a particular value.

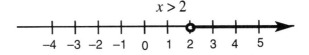

2 is not a solution and should be represented as shown.

The endpoint is a solution if the variable is either (1) less than or equal to, or (2) greater than or equal to, a particular value.

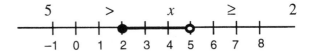

In this case 2 is the solution and should be represented as shown.

If x and y are real numbers then one and only one of the following statements is true.

$x > y$, $x = y$, or $x < y$.

This is the order property of real numbers.

If a, b, and c are real numbers:

1)   If $a < b$ and $b < c$, then $a < c$.

2)   If $a > b$ and $b > c$, then $a > c$.

This is the transitive property of inequalities.

If a, b, and c are real numbers and $a > b$, then $a + c > b + c$ and $a - c > b - c$. This is the addition property of inequality.

Two inequalities are said to have the same **sense** if their signs of inequality point in the same direction.

The sense of an inequality remains the same if both sides are multiplied or divided by the same positive real number.

For the expression $4 > 3$, if we multiply both sides by 5 we will obtain:

$4 \times 5 > 3 \times 5$

$20 > 15$

The sense of the inequality does not change.

The sense of an inequality becomes opposite if each side is multiplied or divided by the same negative real number.

For example, for the same expression $4 > 3$, if we multiply both sides by –5 we would obtain:

$4 \times -5 < 3 \times -5$

$-20 < -15$

The sense of the inequality becomes opposite.

Inequalities that have the same solution set are called **equivalent inequalities**.

Example 1

Solve the inequality $2x + 5 > 9$.

Begin by adding –5 to both sides, so that we get $2x + 5 + (-5) > 9 + (-5)$.

Combining terms we find $2x > 4$, and by dividing each side by 2, $2x \div 2 > 4 \div 2$, we find $x > 2$.

The solution set is

$x = \{x \mid 2x + 5 > 9 \}$

$= \{x \mid x > 2\}$

That is read, "the set of all x, such that x is greater than 2."

Example 2

Solve the inequality $4x + 3 < 6x + 8$.

In order to solve the inequality $4x + 3 < 6x + 8$, we must find all values of x which make it true. Thus, we wish to obtain x alone on one side of the inequality.

Add $-3 - 6x$ to both sides:

$4x + 3 - 3 - 6x < 6x + 8 - 3 - 6x$

$-2x < 5$

In order to obtain x alone we must divide both sides by –2. Recall that dividing an inequality by a negative number reverses the inequality sign, hence

$$\frac{-2x}{-2} > \frac{5}{-2}$$

Cancelling $^{-2}/_{-2}$ we obtain, $x > -\,^{5}/_{2}$.

Thus, our solution is $\{x \mid x > -\,^{5}/_{2}\}$ (the set of all x, such that x is greater than $-\,^{5}/_{2}$).

Questions

Find the solution set for each inequality.

1. $3m + 2 < 7$

2. $\dfrac{1}{2} - 3 \leq 1$

3. $-3p + 1 \geq 16$

4. $-\dfrac{1}{2} < 2r$

5. $\dfrac{2}{3} + 5 \geq 11$

6. $x + 5 < 3x + 3$

7. $-3x + 12 < 11$

8. $11x > 6x - 10$

9. $\dfrac{3}{2} - 9 < -2$

10. $-\dfrac{4}{5} - 1 > 3$

Answers

1. $m < \dfrac{5}{3}$

   $3m + 2 - 2 < 7 - 2$

   $\quad\quad 3m \div 3 < 5 \div 3$

   $\quad\quad\quad\quad m < \dfrac{5}{3}$

2. $x \leq 8$

   $\dfrac{1}{2}x - 3 + 3 \leq 1 + 3$

   $\dfrac{1}{2}x \div \dfrac{1}{2} \leq 4 \div \dfrac{1}{2}$

   $\quad\quad\quad\quad x \leq 8$

3. $p \leq -5$

   $-3p + 1 - 1 \geq 16 - 1$

   $\quad -3p \div -3 \geq 15 \div -3$

   $\quad\quad\quad\quad p \leq -5$

4. $r > -\dfrac{1}{4}$

   $-\dfrac{1}{2} \div 2 < 2r \div 2$

   $\quad\quad -\dfrac{1}{4} < r$

   $\quad\quad\quad r > -\dfrac{1}{4}$

5. $x \geq 9$

   $\dfrac{2}{3}x + 5 - 5 \geq 11 - 5$

   $\dfrac{2}{3}x \div \dfrac{2}{3} \geq 6 \div \dfrac{2}{3}$

   $\quad\quad\quad\quad x \geq 9$

6. $x > 1$

   $x + 5 - x - 3 < 3x + 3 - x - 3$

   $\quad\quad 2 \div 2 < 2x \div 2$

   $\quad\quad\quad\quad 1 < x$

   $\quad\quad\quad\quad x > 1$

7. $x > \dfrac{1}{3}$

   $-3x + 12 - 12 < 11 - 12$

   $\quad -3x \div -3 < -1 \div -3$

   $\quad\quad\quad\quad x > \dfrac{1}{3}$

8. $x > -2$

   $11x - 6x > 6x - 10 - 6x$

   $\quad 5x \div 5 > -10 \div 5$

   $\quad\quad\quad\quad x > -2$

9. $x < \dfrac{14}{3}$

$$\dfrac{3}{2}x - 9 + 9 < -2 + 9$$

$$\dfrac{3}{2}x \div \dfrac{3}{2} < 7 \div \dfrac{3}{2}$$

$$x < \dfrac{14}{3}$$

10. $x < -5$

$$-\dfrac{4}{5}x - 1 + 1 > 3 + 1$$

$$-\dfrac{4}{5}x \div -\dfrac{4}{5} > 4 \div -\dfrac{4}{5}$$

$$x < -5$$

☞ Practice: Algebra

DIRECTIONS: Answer each question as indicated.

Simplify the following expressions.

1. $7a^2c - a + 4a^2c - 4c$

2. $2x - 12y + 8w + 8y - 5x$

Multiply the following monomials and polynomials.

3. $(6pt^2)(-4p^2t^3)$

4. $(7x)(3x - 5y + 9)$

Divide the following monomials and polynomials.

5. $20b^5c^6 \div 2bc^3$

6. $(m^2 + 4m - 15) \div (m - 3)$

Factor the following.

7. $100x^2 - 9y^2$

8. $6w^3 + 9w^2$

9. $x^2 + 9x + 20$

10. $x^2 - 10x + 24$

11. $5x^2 - 22x + 8$

12. $2x^2y^2 - xy - 15$

Solve each equation for the variable given.

13. $8m - 19 = 13$

14. $0.7y + 9 = y + 15$

15. $6c - \dfrac{1}{4} = 2c + \dfrac{1}{4}$

16. $10(x - 3) = x - 36$

17. $3x - 7 < 11x + 1$

18. $\dfrac{1}{6}p + 14 < 20$

19. $2w + \dfrac{1}{4} > \dfrac{1}{3}$

20. $\dfrac{6}{n} = \dfrac{24}{60}$

21. $\dfrac{3}{5} = \dfrac{n}{75}$

22. At a small auto plant, two out of every 25 autos made have the cruise-control feature. If there are a total of 175 autos, how many have cruise-control?

23. In a class of 36 students, there are 28 females. Write as a reduced fraction the ratio of male to female students.

24. A family has a monthly income of $3,200, from which they spend $400 on food and $800 on rent. What is the ratio, in reduced form, of the combined amount spent for food and rent to the total monthly income?

25. Five out of 6 students at a local college take a math course. If 360 students are taking a math course, how many students at this college are not taking a math course?

Answers

1. $11a^2c - 5a$

   $7a^2c - a + 4a^2c - 4a$

   $= (7a^2c + 4a^2c) + (-a - 4a)$

   $= 11a^2c - 5a$

2. $-3x - 4y + 8w$

   $2x - 12y + 8w + 8y - 5x$

   $= (2x - 5x) + (-12y + 8y) + 8w$

   $= -3x - 4y + 8w$

3. $-24p^3t^5$

   $(6pt^2)(-4p^2t^3)$

   $= (6 \times -4)(p \times p^2)(t^2 \times t^3)$

   $= -24p^3t^5$

4. $21x^2 - 35xy + 63x$

   $(7x)(3x - 5y + 9)$

   $= (7 \times 3)(x \times x) + (7 \times -5)(x \times y) + (7 \times 9)(x)$

   $= 21x^2 - 35xy + 63x$

5. $10b^4c^3$

   $20b^5c^6 \div 2bc^3$

   $= (20 \div 2)(b^5 \div b)(c^6 \div c^3)$

   $= 10b^4c^3$

6. $m + 7 + \dfrac{6}{m-3}$

$$\begin{array}{r} m + 7 + \dfrac{6}{m-3} \\ m - 3 \enclose{longdiv}{m^2 + 4m - 15} \\ \underline{m^2 - 3m} \\ 7m - 15 \\ \underline{7m - 21} \\ 6 \end{array}$$

7. $(10x - 3y)(10x + 3y)$

   $100x^2 - 9y^2 = (10x - 3y)(10x + 3y)$

8. $(3w^2)(2w + 3)$

   $6w^3 + 9w^2 = (3w^2)(2w + 3)$

9. $(x + 5)(x + 4)$

   $x^2 + 9x + 20 = (x + 5)(x + 4)$

10. $(x - 6)(x - 4)$

    $x^2 - 10x + 24 = (x - 6)(x - 4)$

11. $(5x - 2)(x - 4)$

    $5x^2 - 22x + 8 = (5x - 2)(x - 4)$

12. $(2xy + 5)(xy - 3)$

    $2x^2y^2 - xy - 15 = (2xy + 5)(xy - 3)$

13. $m = 4$

$$8m - 19 + 19 = 13 + 19$$
$$8m \div 8 = 32 \div 8$$
$$m = 4$$

14. $y = -20$

$$0.7y - y - 9 = y + 15 - y - 9$$
$$-0.3y \div -0.3 = 6 \div -0.3$$
$$y = -20$$

15. $c = \dfrac{1}{8}$

$$6c - \frac{1}{4} - 2c + \frac{1}{4} = 2c + \frac{1}{4} - 2c + \frac{1}{4}$$

$$4c \div 4 = \frac{1}{2} \div 4$$

$$c = \frac{1}{8}$$

16. $x = -\dfrac{2}{3}$

$$10(x - 3) = x - 36$$
$$10x - 30 - x + 30 = x - 36 - x + 30$$
$$9x \div 9 = -6 \div 9$$

$$x = -\frac{2}{3}$$

17. $x > -1$

$$3x - 7 - 3x - 1 < 11x + 1 - 3x - 1$$
$$-8 \div 8 < 8x \div 8$$
$$-1 < x$$
$$x > -1$$

18. $p < 36$

$$\frac{1}{6}p + 14 - 14 < 20 - 14$$

$$\frac{1}{6}p \div \frac{1}{6} < 6 \div \frac{1}{6}$$

$$p < 36$$

19. $w > \dfrac{1}{24}$

$$2w + \frac{1}{4} - \frac{1}{4} > \frac{1}{3} - \frac{1}{4}$$

$$2w \div 2 > \frac{1}{12} \div 2$$

$$w > \frac{1}{24}$$

20. $n = 15$

$$\frac{6}{n} \times n \times 60 = \frac{24}{60} \times n \times 60$$
$$6 \times 60 = n \times 24$$
$$360 \div 24 = 24n \div 24$$
$$15 = n$$

21. $n = 45$

$$\frac{3}{5} \times 75 = \frac{n}{75} \times 75$$
$$45 = n$$

22. 14 autos.

Let n = number of autos with cruise control. Then,

$$\frac{2}{25} \times 175 = \frac{n}{175} \times 175$$
$$14 = n$$

23. $\dfrac{2}{7}$.

28 females implies 8 male students, since $36 - 28 = 8$. The ratio is $\dfrac{8}{28}$, which reduces to $\dfrac{2}{7}$.

24. $\dfrac{3}{8}$.

The combined amount of food and rent is $400 + $800 = $1,200. The ratio becomes $\dfrac{1,200}{3,200}$, which reduces to $\dfrac{3}{8}$.

25. 72 students.

Let n = number of students in the local college. Then $\dfrac{5}{6} = \dfrac{360}{n}$, which yields $n = 432$.

Since 360 students are taking math, then $432 - 360 = 72$ are not taking a math course.

REVIEW

Algebra is the area of mathematics where we begin to represent numbers with symbols in order to learn more about how to use them. There is a lot of new terminology in algebra and it is important to know it all in order to fully understand the concepts. We must recognize and understand how to use terms with variables and coefficients, and how they are used in expressions like monomials or polynomials. We now know how to classify real numbers in sets like positive and negative numbers, and find the absolute value of either, and this will continue to be useful in other kinds of math.

It is important to recall the various operations of monomials and polynomials. For adding and subtracting monomials and polynomials, remember that for like terms with the same variables and exponents, keep the variables the same while adding the coefficients. Multiplication and division of like monomials with exponents requires adding or subtracting (respectively) the exponents while keeping the base the same. With polynomials, the same rules apply, but it is also important to remember the order of operations.

When finding common factors and multiples for a set of numbers, there is a high degree of trial and error. That said, it is worth note that there are a number of common formulas that we should be able to recognize, such as $(a + b)(a - b) = a^2 - b^2$.

For equations and inequalities, it is imperative to remember that any operation done to one side must be done to the other. Know the signs for inequalities: $<$, $>$, $\leq$, $\geq$, and $\neq$, and keep in mind that multiplying or dividing by a negative makes the "greater than" signs switch to less than signs, and vice versa.

Mathematics

Advanced Topics

MATHEMATICS

ADVANCED TOPICS

PROBABILITY

Now that we have a good understanding of mathematics, we want to discuss some other topics that are a little bit more involved. The first of these topics is probability. Geometry, statistics, multi-step word problems, and data analysis will be covered later in this section.

In probability, we are studying the chance of something happening. This is called a **successful outcome**. In other words, we want to know what is the chance that a successful outcome will occur in a certain situation. Here we will be looking at results that are not under our control. That is, the outcome is random. We come across a lot of places where probability can be used. Two examples would be the state lottery and game shows.

The probability of an outcome is expressed between 0 and 1. Percents can also be used here. A probability of 0 (or 0%) denotes an impossibility. An event with 0% probability will never happen. A probability of 1 (or 100%) must happen. An event with 100% probability will always occur. All other probabilities lie somewhere between 0 and 1, that is, between 0% and 100% (if percentages are used). Below is the general formula that can be used to calculate the probability of an event occurring.

$$\text{Probability} = \frac{\substack{\text{Number of ways} \\ \text{a successful outcome} \\ \text{can occur}}}{\substack{\text{Number of possible} \\ \text{outcomes}}}$$

Example 1

Ten pieces of paper are numbered 1 through 10 [1, 2, 3, 4, 5, 6, 7, 8, 9, and 10] and then placed in a box. Without looking, one piece of paper is drawn from the box. Find the probability of the following events occurring.

(a) What is the probability of drawing the 6?

Since only one piece of paper contains a 6, there is only 1 way for this outcome to occur. With ten pieces of paper in the box, each having an equal chance to be drawn, there are 10 possible outcomes that can happen. We can now put this information into our formula and calculate the probability.

$$\text{Probability} = \frac{\substack{\text{Number of ways} \\ \text{a successful outcome} \\ \text{can occur}}}{\substack{\text{Number of possible} \\ \text{outcomes}}}$$

$$\text{Probability} = \frac{1}{10} = 0.10$$

Therefore, the probability of drawing the 6 is 0.1 or 10%.

(b) What is the probability of drawing the 3?

Since only one piece of paper contains a 3, there is only 1 way for this outcome to occur. With ten pieces of paper in the box, each having an equal chance to be drawn, there are 10 possible outcomes that can happen. We can now put this information into our formula and calculate the probability.

$$\text{Probability} = \frac{\text{Number of ways a successful outcome can occur}}{\text{Number of possible outcomes}}$$

$$\text{Probability} = \frac{1}{10} = 0.10$$

Therefore, the probability of drawing the 3 is 0.1 or 10%.

(c) What is the probability of drawing a 4 or less?

We must first determine how many successful outcomes are possible. Since drawing a 1, 2, 3, or 4 would be a success, we know that there are four possible ways for this outcome to occur. With ten pieces of paper in the box, each having an equal chance to be drawn, there are ten possible outcomes that can happen. We can now put this information into our formula and calculate the probability.

$$\text{Probability} = \frac{\text{Number of ways a successful outcome can occur}}{\text{Number of possible outcomes}}$$

$$\text{Probability} = \frac{4}{10} = 0.40$$

Therefore, the probability of drawing a 4 or less is 0.4 or 40%.

(d) What is the probability of drawing an even number?

We must first determine how many successful outcomes are possible. Since drawing a 2, 4, 6, 8, or 10 would be a success, we know that there are five possible ways for this outcome to occur. With ten pieces of paper in the box, each having an equal chance to be drawn, there are ten possible outcomes that can happen. We can now put this information into our formula and calculate the probability.

$$\text{Probability} = \frac{\text{Number of ways a successful outcome can occur}}{\text{Number of possible outcomes}}$$

$$\text{Probability} = \frac{5}{10} = 0.50$$

Therefore, the probability of drawing an even number is 0.5 or 50%.

Example 2

Twenty marbles are placed in a can. Two are black, six white, five red, four blue, and three yellow. Without looking, one marble is drawn from the can. What is the probability of the following occurring?

(a) What is the probability of drawing a blue marble?

Since our can contains four blue marbles, there are four ways for a successful outcome to occur. We have a total of 20 marbles in the can, and each has an equal chance of being drawn. Hence, there are 20 possible outcomes. We can now put this information into our formula and calculate the probability.

$$\text{Probability} = \frac{\text{Number of ways a successful outcome can occur}}{\text{Number of possible outcomes}}$$

$$\text{Probability} = \frac{4}{20} = 0.20$$

Therefore, the probability of drawing a blue marble is 0.2 or 20%.

(b) What is the probability of drawing a black marble?

Since our can contains two black marbles, there are two ways for a successful outcome to occur. We have a total of 20 marbles in the can, and each has an equal chance of being drawn. Hence, there are 20 possible outcomes. We can now put this information into our formula and calculate the probability.

$$\text{Probability} = \frac{\text{Number of ways a successful outcome can occur}}{\text{Number of possible outcomes}}$$

$$\text{Probability} = \frac{2}{20} = 0.10$$

Therefore, the probability of drawing a black marble is 0.1 or 10%.

(c) What is the probability of drawing a white marble?

Since our can contains six white marbles, there are six ways for a successful outcome to occur. We have a total of 20 marbles in the can, and

each has an equal chance of being drawn. Hence, there are 20 possible outcomes. We can now put this information into our formula and calculate the probability.

$$\text{Probability} = \frac{\text{Number of ways a successful outcome can occur}}{\text{Number of possible outcomes}}$$

$$\text{Probability} = \frac{6}{20} = 0.30$$

Therefore, the probability of drawing a white marble is 0.3 or 30%.

(d) What is the probability of drawing a red, white, or blue marble?

Since our can contains five red, six white, and four blue marbles, there are 15 ways for a successful outcome to occur. We have a total of 20 marbles in the can, and each has an equal chance of being drawn. Hence, there are 20 possible outcomes. We can now put this information into our formula and calculate the probability.

$$\text{Probability} = \frac{\text{Number of ways a successful outcome can occur}}{\text{Number of possible outcomes}}$$

$$\text{Probability} = \frac{15}{20} = 0.75$$

Therefore, the probability of drawing a red, white, or blue marble is 0.75 or 75%.

Example 3

A pair of dice are rolled on a table. Find the probability of the following events occurring.

(a) What is the probability of rolling a 1?

Since we are rolling two dice, the smallest value possible is 2 (a 1 on each die). Therefore, there are 0 ways for a successful outcome happening. Since each of the dice contain six sides and each side has an equal chance of coming up, we know that there are $6 \times 6 = 36$ possible outcomes that can occur. Each of the dice accounts for one of the sixes in the equation above. We can now put this information into our formula and calculate the probability.

$$\text{Probability} = \frac{\text{Number of ways a successful outcome can occur}}{\text{Number of possible outcomes}}$$

$$\text{Probability} = \frac{0}{36} = 0$$

Therefore, the probability of rolling a 1 is 0 or 0%.

(b) What is the probability of rolling a 3?

Since we are rolling two dice, we can see that there are two ways for a successful outcome happening. We can roll a 1 on the first die and then a 2 on the second, or a 2 can appear on the first die and a 1 on the second. Since each of the dice contain six sides and each side has an equal chance of coming up, we know that there are $6 \times 6 = 36$ possible outcomes that can occur. Each of the

dice account for one of the sixes in the equation above. We can now put this information into our formula and calculate the probability.

$$\text{Probability} = \frac{\text{Number of ways a successful outcome can occur}}{\text{Number of possible outcomes}}$$

$$\text{Probability} = \frac{2}{36} = 0.05$$

Therefore, the probability of rolling a 3 is 0.05 or 5.5%.

(c) What is the probability of rolling an 8?

Since we are rolling two dice, we can see that there are five ways for a successful outcome happening. We can roll a 2 on the first die and then a 6 on the second; a 6 can appear on the first die and a 2 on the second; a 3 on the first die and a 5 on the second; a 5 on the first die and a 3 on the second; or a 4 on both dice. Since each of the dice contain six sides and each side has an equal chance of coming up, we know that there are $6 \times 6 = 36$ possible outcomes that can occur. Each of the dice account for one of the sixes in the equation above. We can now put this information into our formula and calculate the probability.

$$\text{Probability} = \frac{\text{Number of ways a successful outcome can occur}}{\text{Number of possible outcomes}}$$

$$\text{Probability} = \frac{5}{36} = 0.138$$

Therefore, the probability of rolling an 8 is 0.138 or 13.8%.

(d) What is the probability of rolling a 10?

Since we are rolling two dice, we can see that there are three ways for a successful outcome happening. We can roll a 5 on each dice; a 6 on the first and a 4 on the second; or a 4 on the first and a 6 on the second. Since each of the dice contain six sides, and each side has an equal chance of coming up, we know that there are 6 × 6 = 36 possible outcomes that can occur. Each of the dice account for one of the sixes in the equation above. We can now put this information into our formula and calculate the probability.

$$\text{Probability} = \frac{\text{Number of ways a successful outcome can occur}}{\text{Number of possible outcomes}}$$

$$\text{Probability} = \frac{3}{36} = 0.08\overline{3}$$

Therefore, the probability of rolling a 10 is 0.083 or 8.3%.

Questions

Five pieces of paper are numbered 1 through 5 [1, 2, 3, 4, and 5] and then placed in a box. Without looking, one piece of paper is drawn from the box. Find the probability of the following events occurring.

1.   What is the probability of drawing a 2?

2.   What is the probability of drawing a 5?

3.   What is the probability of drawing a 1 or 2?

4.   What is the probability of drawing an odd number?

5.   What is the probability of drawing a 3 or greater?

Fifteen marbles are placed in a can. Five are pink, six blue, two yellow, one orange, and one white. Without looking, one marble is drawn from the can. Find the probability of the following events occurring.

6.   What is the probability of drawing an orange marble?

7.   What is the probability of drawing a blue marble?

8.   What is the probability of drawing a yellow or white marble?

9.   What is the probability of drawing a pink marble?

10.   What is the probability of drawing a pink, yellow, orange, or white marble?

A pair of dice are rolled on a table. Find the probability of the following events occurring.

11.   What is the probability of rolling a 5?

12.   What is the probability of rolling a 7?

13.   What is the probability of rolling a 12?

14.   What is the probability of rolling an 11?

15.   What is the probability of rolling a number less than 13?

Answers

1. 0.2 or 20%

    Probability of drawing a 2

    $$= \frac{1}{5} = 20\%$$

2. 0.2 or 20%

    Probability of drawing a 5

    $$= \frac{1}{5} = 20\%$$

3. 0.4 or 40%

    Probability of drawing a 1 or 2

    $$= \frac{2}{5} = 40\%$$

4. 0.6 or 60%

    Probability of drawing an odd number equals the probability of drawing a 1, 3, or 5

    $$= \frac{3}{5} = 60\%$$

5. 0.6 or 60%

    Probability of drawing a 3 or greater equals the probability of drawing a 3, 4, or 5

    $$= \frac{3}{5} = 60\%$$

6. 6.66%

    Probability of drawing orange

    $$= \frac{1}{15} = 6.66...\%$$

7. 0.4 or 40%

    Probability of drawing blue

    $$= \frac{6}{15} = 40\%$$

8. 0.2 or 20%

    Probability of drawing yellow or white

    $$= \frac{2+1}{15} = \frac{3}{15} = 20\%$$

9. 0.3... (or 33.33...%)

    Probability of drawing pink

    $$= \frac{5}{15} = 33.33...\%$$

10. 0.6 or 60%

    Probability of drawing pink, yellow, orange, or white

    $$= \frac{5+2+1+1}{15} = \frac{9}{15} = 60\%$$

11. 0.11... (or 11.11...%)

    There are 36 total outcomes. The four ways to roll a 5 are: 1, 4; 4, 1; 2, 3; 3, 2. The required probability is

    $$\frac{4}{36} = 11.11...\%$$

12. 0.166... (or 16.66...%)

    The six ways to roll a 7 are 1, 6; 6, 1; 2, 5; 5, 2; 3, 4; 4, 3. Since there are 36 possible outcomes, the required probability is

    $$\frac{6}{36} = 16.66...\%$$

13. 0.0277... (or 2.77...%)

    The only way to roll a 12 is 6, 6. With 36 possible outcomes, the required probability is

$$\frac{1}{36} = 2.77...\%$$

14. 0.055... (or 5.55...%)

The two ways to roll an 11 are 5, 6 and 6, 5. With 36 possible outcomes, the required probability is

$$\frac{2}{36} = 5.55...\%$$

15. 1 or 100%

No matter how the dice are rolled, the number shown must be between 2 and 12 inclusive. Thus the probability of rolling a number less than 13 is a certainty, which is 1.

STATISTICS

Statistics involves assembling, organizing, and analyzing data, as well as drawing conclusions about what that data means. We encounter statistics on a daily basis. Stock market reports, public opinion polls, and even the past performances of athletes are generally presented to us in statistical form. Advertisers also try to convince us their products are the best by presenting us with basic statistical information.

When dealing with statistics, we usually have to work with basic mathematical operations such as addition, subtraction, multiplication, and division. The following examples will show this in more detail.

Example 1

Find the arithmetic mean of the numbers 3, 7, 1, 24, 11, and 32.

> The arithmetic mean, or mean, of a set of measurements is the sum of the measurements divided by the total number of measurements.

The arithmetic mean of a set of numbers is the same as the average of a set, which you learned about in the Topics in Mathematics section. However, here it is presented in a more sophisticated way.

The arithmetic mean of a set of numbers x_1, x_2 ..., x_n is denoted by $\bar{x}$ (read "x bar").

$$\bar{x} = \frac{\sum_{i=1}^{n} x_1}{n} = \frac{x_1 + x_2 + ... + x_n}{n}$$

$$= \frac{3 + 7 + 1 + 24 + 11 + 32}{6} = 13$$

Example 2

Find the mode of the set of numbers 2, 2, 4, 7, 9, 9, 13, 13, 13, 26, and 29.

> The **mode** of a set of numbers is the number that appears most frequently. since 13 appears 3 times, while no other number appears more than twice, it is the mode.

> The set of numbers that has two or more modes is called **bimodal**.

> For grouped data—data presented in the form of a frequency table—we do not know the actual measurements, only how many measurements fall into each interval. In such a case the mode is the midpoint of the class interval with the highest frequency.

> Note that the mode can also measure popularity. In this sense, we can determine the most popular model of car or the most popular actor.

Example 3

Find the median of the following sets of numbers.

(a) The scores of a test are 78, 79, 80, 83, 87, 92, and 95.

The **median** is the middle number in a sequence. Hence the median is 83.

(b) 21, 25, 29, 33, 44, and 47.

It is more difficult to compute the median for grouped data. The exact value of the measurements is not known; hence, we know only that the median is located in a particular class interval. The problem is where to place the median within this interval. In this special case, we find the mean of the two middle numbers, so that

$$\frac{29 + 33}{2} = 31.$$

Questions

If needed, round your answers to the hundredths place.

1. The following measurements were taken by an antique dealer as he weighed to the nearest pound his prize collection of anvils. The weights were 84, 92, 37, 50, 50, 84, 40, and 98. What was the mean weight of the anvils?

2. The numbers 4, 2, 7, and 9 occur with frequencies 2, 3, 11, and 4, respectively. Find the arithmetic mean.

3. The pilot has to pass three tests. The second test is weighted three times as much as the first, and the third test is weighted four times as much as the first. The pilot researched the score of 40 on the first test, 45 on the second test, and 60 on the third test. What is the weighted mean?

4. A student takes two quizzes, one midterm, and one final exam in a statistics course. The midterm counts three times as much as a quiz, and the final exam counts five times as much as a quiz. If the quiz scores were 70 and 80, the midterm score was 65, and the final exam score was 85, what was the weighted average?

Find the mode, or modes, of the following samples.

5. 14, 19, 16, 21, 18, 24, 15, and 19.

6. 6, 7, 7, 3, 8, 5, 3, and 9.

7. 14, 16, 21, 19, 18, 24, and 17.

Find the median of the following samples.

8. 34, 29, 26, 37, and 31.

9. 34, 29, 26, 37, 31, and 34.

10. If the measurements 3, 7, 2, 8, 0, and 4 occur with the frequencies 3, 2, 1, 5, 10, and 6 respectively, what is the arithmetic mean?

Answers

1. 66.88 pounds

The average or mean weight of the anvils is

$$\bar{x} = \frac{\text{sum of the observations}}{\text{number of observations}}$$

$$= \frac{84 + 92 + 37 + 50 + 50 + 84 + 40 + 98}{8}$$

$$= 535/8$$

$$= 66.88$$

2. 6.35

To find the arithmetic mean, *x*, multiply each different number by its associated frequency. Add these products, then divide by the total number of numbers.

$$\bar{x} = \frac{[(4)(2) + (2)(3) + (7)(11) + (9)(4)]}{2 + 3 + 11 + 4}$$

$$= (8 + 6 + 77 + 36) \div 20$$

$$= 127 \div 20 = 6.35$$

3. 51.88

The weighted mean is

$$\frac{40 \times 1 + 45 \times 3 + 60 \times 4}{8} = 51.88$$

4. 77

$$\bar{x} = [(70)(1) + (80)(1) + (65)(3) + (85)(5)] \div (1 + 1 + 3 + 5)$$

$$= (70 + 80 + 195 + 425) \div 10$$

$$= 770 \div 10 = 77$$

5. 19

The number 19 is observed three times in this sample, and no other number appears as frequently. The mode of this sample is therefore 19.

6. 3 and 7

In this sample the numbers 7 and 3 both appear twice. There are no other numbers that appear as frequently as these two. Therefore, 3 and 7 are the modes of this sample.

7. No modes.

In this sample all the numbers occur with the same frequency. There is no single number that is observed more frequently than any other. Thus, there is no mode.

8. 31

The median is the middle number. The number of observations that lie above the median is the same as the number of observations that lie below it. Arranged in order we have 26,

29, 31, 34, and 37. The number of observations is odd, and thus the median is 31.

9. 32.5

The sample arranged in order is 26, 29, 31, 34, 34, and 37. The number of observations is even and thus the median, or middle number, is chosen halfway between the third and fourth numbers. In this case, the median is

$$\frac{31 + 34}{2} = 32.5$$

10. 3.30

The arithmetic mean is

$$\bar{x} = \frac{3 \times 3 + 7 \times 2 + 2 \times 1 + 8 \times 5 + 0 \times 10 + 4 \times 6}{3 + 2 + 1 + 5 + 10 + 6}$$

$$= 3.30$$

GEOMETRY

Geometry is not a new topic for us. We have already introduced some ideas of geometry, but at that time we were looking at measurements. Squares, rectangles, and boxes were discussed in that section. We now want to consider triangles and circles.

A **triangle** is a three-sided object with three interior angles. The interior angles add up to 180 degrees. All angles throughout this book will be measured in degrees. Below are a few examples of triangles.

Triangles can be classified in a number of different ways. Most of these depend on the angle measures. Below are a few of these classifications with their definitions.

In a *right triangle* one of the interior angles has a measure of 90 degrees. This is normally denoted by placing a small square in that angle's position. The first triangle in the box on page 210 is a right triangle.

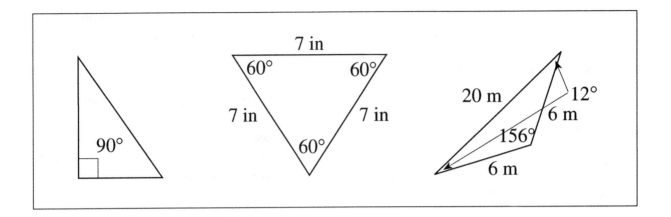

An *acute triangle* is one where all the angle measures are less than 90 degrees. The angles can be the same or different measures, but each must be smaller than 90 degrees. The second triangle in the box above is an acute triangle.

In an *equilateral triangle* all of the sides are the same length and each of the interior angles has a measure of 60 degrees. This type of triangle is a special case of an acute triangle. The second triangle in the box above is also an equilateral triangle.

An *isosceles triangle* has two sides with equal measure and two angles with the same measure. The equal sides are opposite the equal angles. The third triangle in the box above is an isosceles triangle.

In an *obtuse triangle* one of the interior angles has a measure of greater than 90 degrees. Since the total angle measure must be 180 degrees, when one angle is larger than 90 degrees, the others must be smaller. The third triangle in the box above is also an obtuse triangle.

Example 1

Write the term (or terms) that describe the following triangles. Use right, acute, equilateral, isosceles, or obtuse. Some of the triangles may be described using more than one term.

(a) Since each interior angle is less than 90 degrees, we know that this is an acute triangle.

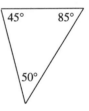

(b) Since one of the interior angles of this triangle has a measure of 90 degrees, we know that this is a right triangle.

(c) Since each interior angle is less than 90 degrees, we know that this is an acute triangle. Also, since we have two angles that are equal and the sides opposite them are equal, we know that it is also an isosceles triangle.

(d) Since one of the interior angles of this triangle is greater than 90 degrees, we know that it is an obtuse triangle.

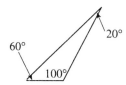

As with squares and rectangles, we can find the perimeter of a triangle. Recall that the perimeter is the distance around the outside of the object. The only problem is that we must be given the distance of each side. Below we will consider a few examples.

Example 2

Find the perimeter of the following triangles.

(a) The perimeter is the distance around the outside of the object. Therefore, we must add the lengths of the sides together.

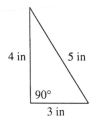

Perimeter = (Side 1) + (Side 2) + (Side 3)

Perimeter = (3) + (4) + (5) = 12 inches

Thus, the perimeter is 12 inches.

(b) This triangle is equilateral. We know that all the sides are the same length. Since we know that one side has a measure of 3 meters, we know that all sides have that measure.

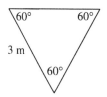

Perimeter = (Side 1) + (Side 2) + (Side 3)

Perimeter = (3) + (3) + (3) = 9 meters

Thus, the perimeter is 9 meters.

(c) This triangle is isosceles. We know that the two sides opposite the equal angles are also equal. Therefore, the other side also has a measure of 6 feet.

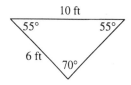

Perimeter = (Side 1) + (Side 2) + (Side 3)

Perimeter = (6) + (6) + (10) = 22 feet

Thus, the perimeter is 22 feet.

(d) This triangle is obtuse. Since we are only given the measurement on two sides and have no way of finding the length of the other sides, we cannot answer this question. Notice that this is not one of our special cases where we can deduce the length of the other side.

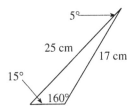

A **circle** is an object where every point on the curve is an equal distance from a center point. We now need to introduce some terminology that is particular to circles. If we draw a straight line from one side of the circle, through the center, and then continue to the other side, this line is called the *diameter*.

The *radius* is half of the diameter. In this case, we measure the line from one side of the circle to the center. Earlier, we discussed the perimeter of an object. This is the distance around the outside of the object. In a circle, this is called the *circumference*.

All circles have a set ratio between the circumference and the diameter. This ratio is represented by the Greek letter pi. This value is approximately 3.14 (rounded to two decimal positions). It can also be approximated as a fraction as 22/7, but this is not the exact value. For us, we will use 3.14 as pi. The formula for the circumference is

$$\text{Circumference} = (\text{pi}) \times (\text{diameter})$$

Example 3

Find the circumference of the following circles. If needed, round your answer to the hundredths place.

(a) Circle with diameter of 16 inches.

Since we know the diameter of this circle, we can calculate the circumference easily.

Circumference = (pi) × (diameter)

Circumference = (3.14) × (16)
= 50.24 inches

Therefore, the circumference is 50.24 inches.

(b) Circle with diameter of 2 feet.

Since we know the diameter of this circle, we can calculate the circumference easily.

Circumference = (pi) × (diameter)

Circumference = (3.14) × (2)
= 6.28 feet

Therefore, the circumference is 6.28 feet.

(c) Circle with diameter of 7.3 meters.

Since we know the diameter of this circle, we can calculate the circumference easily.

Circumference = (pi) × (diameter)

Circumference = (3.14) × (7.3)
= 22.922 meters

Therefore, the circumference is 22.92 meters.

(d) Circle with radius of 5.07 inches.

In this circle, we are given the radius. In order to calculate the circumference, we must find the diameter. We know that the radius is half of the diameter. Therefore, the diameter in this example is 10.14 inches (5.07 × 2). Now we are able to calculate the circumference.

Circumference = (pi) × (diameter)

Circumference = (3.14) × (10.14)
= 31.8396 inches

Therefore, the circumference is 31.84 inches.

We now want to find the area of a circle. As with the circumference, pi also plays an important role in this calculation. The formula below can be used to find the area of any circle. Again, we will use 3.14 as pi.

$$\text{Area} = (\text{pi}) \times (\text{radius})^2$$

Example 4

Find the area of the following circles. If needed, round your answer to the hundredths place.

(a) Circle with radius of 3 inches.

Since we know the radius, we can plug its value into our formula and calculate the area.

$$\text{Area} = (\text{pi}) \times (\text{radius})^2$$

$$\text{Area} = (3.14) \times (3)^2$$

$$\text{Area} = (3.14) \times (9) = 28.26 \text{ inches}^2$$

Therefore, the area of this circle is 28.26 square inches.

(b) Circle with radius of 1.6 feet.

Since we know the radius, we can plug its value into our formula and calculate the area.

$$\text{Area} = (\text{pi}) \times (\text{radius})^2$$

$$\text{Area} = (3.14) \times (1.6)^2$$

$$\text{Area} = (3.14) \times (2.56)$$
$$= 8.0384 \text{ feet}^2$$

Therefore, the area of this circle is 8.04 square feet.

(c) Circle with radius of 5 meters.

Since we know the radius, we can plug its value into our formula and calculate the area.

$$\text{Area} = (\text{pi}) \times (\text{radius})^2$$

$$\text{Area} = (3.14) \times (5)^2$$

$$\text{Area} = (3.14) \times (25) = 78.5 \text{ meters}^2$$

Therefore, the area of this circle is 78.5 square meters.

(d) Circle with diameter of 14.2 centimeters.

In this example, we are given the diameter. Since the radius is used in this formula, we must first calculate it. We know that the radius is half of the diameter. Therefore, in this case, the radius is 7.1 centimeters. Now we can plug this value into our formula and find the area.

$$\text{Area} = (\text{pi}) \times (\text{radius})^2$$

$$\text{Area} = (3.14) \times (7.1)^2$$

$$\text{Area} = (3.14) \times (50.41)$$

$$= 158.2874 \text{ centimeters}^2$$

Therefore, the area of this circle is 158.29 square centimeters.

Questions

Write the term (or terms) that describe the following triangles. Use right, acute, equilateral, isosceles, or obtuse. Some of the triangles may be described using more than one term.

1.

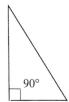

90°

2.

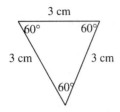

7.

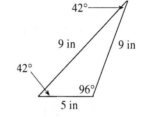

3.

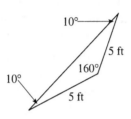

8.

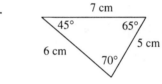

9.

4.

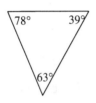

10.

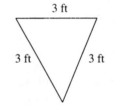

5.

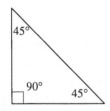

Find the perimeter of the following triangles.

6.

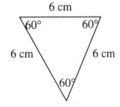

Find the circumference of the following circles. If needed, round your answer to the hundredths place.

11. Circle with diameter of 13 inches.

12. Circle with diameter of 1.5 feet.

13. Circle with diameter of 5 centimeters.

14. Circle with radius of 1.75 inches.

15. Circle with radius of 3 meters.

Find the area of the following circles. If needed, round your answer to the hundredths place.

16. Circle with radius of 7 feet.

17. Circle with radius of 12.6 inches.

18. Circle with radius of 2.25 centimeters.

19. Circle with diameter of 20 meters.

20. Circle with diameter of 36.47 inches.

Answers

1. Right and acute triangle

2. Acute and equilateral triangle

3. Obtuse and isosceles triangle

4. Acute triangle

5. Right and isosceles triangle

6. 18 meters

$$6 \times 3 = 18$$

7. 24 feet

$$8 + 10 + 6 = 24$$

8. 23 inches

$$9 + 9 + 5 = 23$$

9. 18 centimeters

$$6 + 5 + 7 = 18$$

10. 9 feet

$$3 + 3 + 3 = 9$$

11. 40.82 inches

$$Circumference = 3.14 \times 13$$
$$= 40.82 \text{ inches}$$

12. 4.71 feet

$$Circumference = 3.14 \times 1.5$$
$$= 4.17 \text{ feet}$$

13. 15.7 centimeters

$$Circumference = 3.14 \times 5$$
$$= 15.7 \text{ centimeters}$$

14. 10.99 inches

$$Circumference = 2 \times 3.14 \times 1.75$$
$$= 10.99 \text{ inches}$$

15. 18.84 meters

$$Circumference = 2 \times 3.14 \times 3$$
$$= 18.84 \text{ meters}$$

16. 153.86 feet^2

$$Area = 3.14 \times 7^2 = 153.86 \text{ feet}^2$$

17. 498.51 inches^2

$$Area = 3.14 \times 12.6^2$$
$$= 498.51 \text{ inches}^2$$

18. 15.90 centimeters^2

$$Area = 3.14 \times 2.25^2$$
$$= 15.90 \text{ centimeters}^2$$

19. 314 meters^2

$$Radius = \frac{20}{2} = 10.$$

$$Area = 3.14 \times 10^2 = 314 \text{ meters}^2$$

20. 1,044.10 inches^2

$$Radius = \frac{36.47}{2} = 18.235.$$

Area = 3.14×18.235^2
= 1044.10 inches^2

GRAPHS

Coordinate geometry refers to the study of geometric figures using algebraic principles.

The graph below is called the Cartesian coordinate plane. The graph consists of a pair of perpendicular lines called **coordinate axes**. The vertical axis is the **y-axis** and the horizontal axis is the **x-axis**. The point of intersection of these two axes is called the **origin**; it is the zero point of both axes. Furthermore, points to the right of the origin on the x-axis and above the origin on the y-axis represent positive real numbers. Points to the left of the origin on the x-axis or below the origin on the y-axis represent negative real numbers.

The four regions cut off by the coordinate axes are, in counterclockwise direction from the top right, called the first, second, third, and fourth quadrant, respectively. The first quadrant contains all points with two positive coordinates.

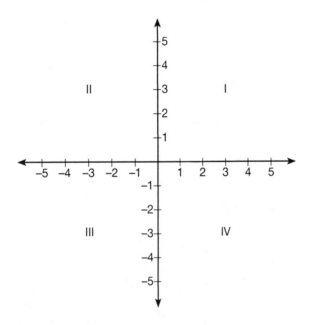

In the graph shown below, two points are identified by the ordered pair (x, y) of numbers. The x-coordinate is the first number and the y-coordinate is the second number.

To plot a point on the graph when given the coordinates, draw perpendicular lines from the number-line coordinates to the point where the two lines intersect.

To find the coordinates of a given point on the graph, draw perpendicular lines from the point to the coordinates on the number line. The x-coordinate is written before the y-coordinate and a comma is used to separate the two.

In this case, point A has the coordinates (4, 2) and the coordinates of point B are (–3, –5).

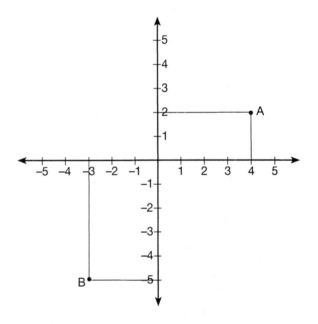

Graphing simple linear equations involves equations like

$$y = mx, \ y = b, \text{ or } x = b$$

where m and b are real numbers. Equations like y = b and x = b represent lines parallel to either the x- or y-axis.

Example 1

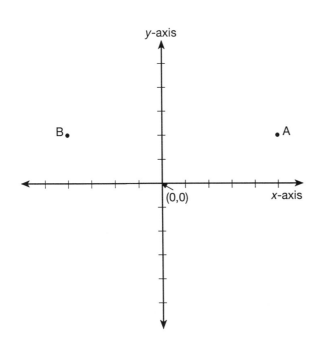

(a) What is the coordinate of point A?

Starting at (0,0), we move 5 units in the positive x direction and 2 units in the positive y direction. The coordinates of point A are (5, 2).

(b) How many units separate point A and point B?

Point B's coordinates are (–4, 2). Since points A and B have the same y coordinates, the distance between them is the absolute value of the difference of their x coordinates, which is $|5 - (-4)|$ or $|-4 - 5| = 9$.

Example 2

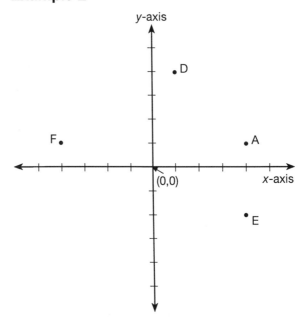

Which other point has the same y coordinate as point C?

Point C is located at (4, 1), point D at (1, 4), point E at (4, –2), and point F at (–4, 1). Thus point F has the same y coordinate as point C.

Example 3

Graph $y = 3$

Since x does not appear in the equation, the value of x has no effect on y. To satisfy the equation y must always be 3.

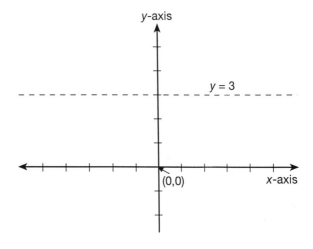

Example 4

Graph $y = 4x$.

Pick some values for x or y and substitute them in the equation to solve for the other variable. Try $y = 0$ and $y = 2$.

$0 = 4x$

$0 = x$

Therefore, $(0, 0)$ satisfies the equation.

$2 = 4x$

$^1/_2 = x$

$(^1/_2, 2)$ is another ordered pair which satisfies the equation.

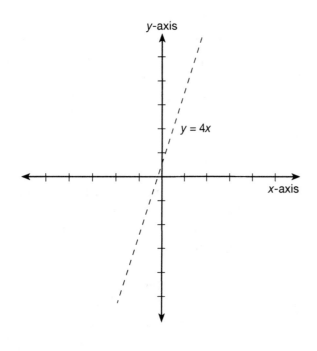

Questions

For questions 1–5, use the following graph, given 5 points.

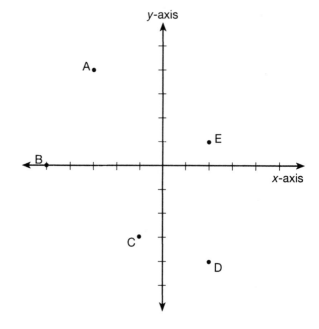

1. What are the coordinates of point D?

2. What are the coordinates of point A?

3. Which point does *not* lie in a quadrant?

4. What is the x-coordinate for point C?

5. How many units apart are points D and E?

For questions 6–10, use the following graph, given 7 points.

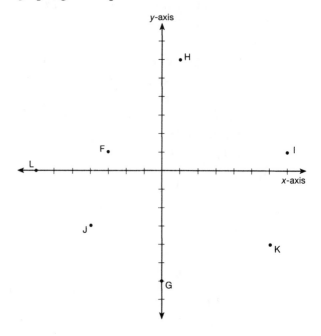

6. How many of these points have positive *x*-coordinates?

7. How many points do *not* lie in Quadrant 1?

8 How many units apart are points F and I?

9. What is the *y*-coordinate of point K?

10. If you start at point J and move 4 units in the positive *x* direction and 6 units in the positive *y* direction, what would be the coordinates of your final position?

Answers

1. Point D is 2 units to the right and 4 units down from (0, 0). Thus, its coordinates are (2, –4).

2. Point A is 3 units to the left and 4 units up from (0, 0). Thus, its coordinates are (–3, 4).

3. The coordinates of point B are (–5, 0). Since it lies on the *x*-axis, it is not located in any quadrant.

4. Point C lies 1 unit to the left and 3 units down from (0, 0). Thus, its *x*-coordinate is –1.

5. Point E is located at (2, 1) and point D is located at (2, –4). The distance between them is 1 – (–4) = 5 units.

6. Points H (1,5), I (7, 1), and K (6, –4) all have positive *x*-coordinates.

7. Points F, L, J, G, K all lie outside Quadrant 1, which is in the upper right-hand corner of the graph.

8. Since F is located at (–2, 1) and I is located at (7, 1), the distance between them is 7 – (–2) = 9.

9. Point K is located at (6, –4).

10. Point J is located at (–4, –3). By moving 4 units in the positive x direction and 6 units in the positive y direction, the new location can be represented by (–4 + 4, –3 + 6) = (0,3).

MULTI-STEP WORD PROBLEMS

Since we are now experts in mathematics, we want to discuss multi-step word problems. Throughout this section, we will be building upon ideas that we have already covered. In these examples, it is best to break the problem into different steps and then solve each of them separately.

The biggest area where this can be seen is when we spend money. Sometimes this is called Consumer Mathematics. A common question is how much money will we have left? To answer this question we must first know how much we have and how much we spent. The following examples will show this in more detail.

Example 1

Solve the following problem.

(a) Kelli goes to the store to get a few groceries. She buys two loaves of bread for $2.19, some oranges for $1.99, a pound of ham for $3.99, and three candy bars for $1.70. If Kelli gives the clerk $10, how much change will she get back?

First we must find out how much she spent. This can be done by adding everything together.

Spent = (Bread) + (Oranges) + (Ham) + (Candy)

Spent = (2.19) + (1.99) + (3.99) + (1.70)

Spent = $9.87

Now that we know that she spent $9.87, we can subtract this value from the amount she gave the clerk to find out how much change she will get back.

Change = (Total) – (Spent)

Change = 10 – 9.87

Change = $0.13

Therefore, she will have $0.13 (or 13 cents) left after her purchases.

(b) George is buying furniture for his new home. He buys a couch and chair for $599, a television for $279, a bed for $978, and a table with chairs for $1,200. If the sales tax is $155 and George gives the sales person $3,500, how much change will he get back?

First we must find out how much he spent. This can be done by adding everything together.

Spent = (Couch & Chair) + (Television) + (Bed) + (Table & Chairs) + Tax

Spent = (599) + (279) + (978) + (1,200) + 155

Spent = $3,211

Now that we know that he spent $3,211, we can subtract this value from the amount he gave the salesperson to find out how much change he will get back.

Change = (Total) – (Spent)

Change = 3,500 – 3,211

Change = $289

Therefore, he will have $289 left after his purchases.

(c) Brian is repairing a car for a client. He purchases four tires for $249, the parts for a brake job for $319, and a fender for $700. If he charges $200 for his work and his client gives him $1,500, how much change does the client get back?

First we must find out how much his client owes. This can be done by adding everything together.

Owe = (Tires) + (Brakes) + (Fender) + (Work)

Owe = (249) + (319) + (700) + (200)

Owe = $1,468

Now that we know that his client owes $1,468, we can subtract this value for the amount of change that the client will get back.

Change = (Total) – (Owed)

Change = 1,500 – 1,468

Change = $32

Therefore, his client will receive $32 back in change.

(d) Dale is working on his bathroom. If he has $250 to spend and purchases plastic piping for $4.99, shut-offs for $39.95, a sink for $99.87, a sink fixture for $29.99, and sales tax is $15.36, how much money will he have left?

First we must find out how much he spent. This can be done by adding everything together.

Spent = (Piping) + (Shut-offs) + (Sink) + (Fixture) + (Tax)

Spent = (4.99) + (39.95) + (99.87) + (29.99) + (15.36)

Spent = $190.16

Now that we know that he spent $190.16, we can subtract this value for the amount of money that he will have left.

Change = (Total) – (Spent)

Change = 250 – 190.16

Change = $59.84

Therefore, he will have $59.84 left after his work.

Solving increase and decrease problems is a good example of how we can split examples into pieces. In most cases, percents are involved. We are looking to see what the final amount will be if the beginning amount is increased (or decreased) by a certain percent. The examples below will show this in more detail.

Example 2

In the following problems, find the amount of increase and the new amount.

(a) The price of a $30 book is increased by 10%. What will be the new price of the book?

In this problem, the increase will be 10% of the original price (or $30). By converting the percent into a decimal and multiplying, we can find the amount of the increase.

Increase = 10% of $30

Increase = $0.10 \times 30 = \$3$

By combining the increase with the original price, we can find the new

price of the book easily. Therefore, the new price will be

(New Price) = (Old Price) + (Increase)

(New Price) = $30 + $3 = $33

Therefore, the new price will be $33.

(b) The price of a television increases by 6%. If the original price is $150, what will be the new price?

In this problem, the increase will be 6% of the original price (or $150). By converting the percent into a decimal and multiplying, we can find the amount of the increase.

Increase = 6% of $150

Increase = $0.06 \times 150 = \$9$

By combining the increase with the original price, we can find the new price of the book easily. Therefore, the new price will be

(New Price) = (Old Price) + (Increase)

(New Price) = $150 + $9 = $159

Therefore, the new price will be $159.

(c) The attendance at a football game increases by 37%. If 12,000 people normally attend, how many will be attending now?

In this problem, the increase will be 37% of the original attendance (or 12,000). By converting the percent into a decimal and multiplying, we can find the amount of the increase.

Increase = 37% of 12,000

Increase = 0.37 × 12,000 = 4,400 people

By combining the increase with the original attendance, we can find the new attendance at the football game easily. Therefore, the new attendance will be

(New Attendance)
= (Old Attendance) + (Increase)

(New Attendance) = 12,000 + 4,400
= 16,400 people

Therefore, the new attendance will be 16,400 people.

(d) A magazine normally contains 25 pages. If the number of pages increases by 145%, how many pages does it now have?

In this problem, the increase will be 145% of the original number of pages (or 25). By converting the percent into a decimal and multiplying, we can find the amount of the increase.

Increase = 145% of 25

Increase = 1.45 × 25 = 36.25 pages

By combining the increase with the original number of pages, we can find the new number of pages in the magazine easily. Therefore, the new number of pages will be

(New Number of Pages)
= (Old Number of Pages)
+ (Increase)

(New Number of Pages)
= 25 + 36.25 = 61.25

Therefore, the magazine will now have approximately 62 pages.

Example 3

In the following problems, find the amount of decrease and the new amount.

(a) A grocery store is having a sale on hot dogs (normally priced at $1.09). If during the sale, the price is reduced by 83%, what is the sale price?

In this problem, the discount will be 83% of the original price (or $1.09). By converting the percent into a decimal and multiplying, we can find the amount of the discount.

Discount = 83% of $1.09

Discount = 0.83 × 1.09 = $0.9047

By combining the discount with the original price, we can find the new price of the hot dogs easily. Therefore, the new price will be

(New Price) = (Old Price)
– (Discount)

(New Price) = $1.09 – $0.9047
= $0.1853

Therefore, during the sale a package of hot dogs will sell for $0.19 (or 19 cents).

(b) A store is going out of business and is selling everything at 55% off. If a desk normally sells for $199, how much will it cost now?

In this problem, the discount will be 55% of the original price (or $199). By converting the percent into a decimal and multiplying, we can find the amount of the discount.

Discount = 55% of $199

Discount = 0.55 × 199 = $109.45

By combining the discount with the original price, we can find the new price of the desk easily. Therefore, the new price will be

(New Price) = (Old Price) – (Discount)

(New Price) = $199 – $109.45 = $89.55

Therefore, during the sale the desk will sell for $89.55.

(c) The attendance at a high school basketball game decreased by 17%. If 125 people normally attend, how many people will attend now?

In this problem, the decrease will be 17% of the original attendance (or 125). By converting the percent into a decimal and multiplying, we can find the amount of the decrease.

Decrease = 17% of 125

Decrease = 0.17 × 125 = 21.25 people

By combining the decrease with the original attendance, we can find the new attendance easily. Therefore, the new attendance will be

(New Attendance) = (Old Attendance) – (Decrease)

(New Attendance) = 125 – 21.25 = 103.75

Therefore, the new attendance at the basketball game will be approximately 104 people.

(d) The weight of a package of cookies decreases by 26%. What is the weight of a new package if the old packages weighed 16 ounces?

In this problem, the decrease will be 26% of the original weight (or 16). By converting the percent into a decimal and multiplying, we can find the amount of the decrease.

Decrease = 26% of 16

Decrease = 0.26 × 16 = 4.16 ounces

By combining the decrease with the original weight, we can find the new weight easily. Therefore, the weight will become

(New Weight) = (Old Weight) – (Decrease)

(New Weight) = 16 – 4.16 = 11.84

Therefore, the new weight of the cookie package will be approximately 12 ounces.

Questions

Solve the following problems.

1. Sam goes to the store for groceries. He buys a turkey for $19.75, bread for $1.99, and peaches for $0.36. If he gives the cashier $30, how much will he get back?

2. Nancy is shopping at a craft store. If she has $10 and purchases styrofoam for $2.99, yarn for $1.99, glue for $0.99, and the sales tax is $0.36, how much will she get back?

3. Edith needs to send some flower arrangements. If she has $400 to spend and sends six arrangements at $39.90 each, how much will she have left?

4. Jim is looking for some new furniture. He is considering a couch/bed for $1,999, a computer for $2,197, and a computer desk for $219. If he has $4,500 and the sales

tax is $75, how much money will he have left?

5. Beth is considering moving into an apartment. The rent is $599, electricity is $179, water is $52, and her car payment is $263. If she earns $1,500 a month, how much money will she have left to use for entertaining?

In the following problems, find the amount of increase and the new amount.

6. The price of a newspaper increases by 7%. If it originally cost $1.25, what would be the new price?

7. The price of a car increases by 16.2%. If it originally cost $13,000, what would be the new price?

8. The attendance at a hockey game increases by 58%. If the original attendance is 37,000, what would be the new attendance?

9. The population of a town increased by 201%. If the original population was 47,123, what would be the new population?

10. The number of books published by REA increases by 2.572%. If they originally published 150 books, how many books do they now publish?

In the following problems find the amount of decrease and the new amount.

11. A department store is having a sale on radios. A deluxe model that originally sold for $219 is on sale for 35% off. What is the sale price of this radio?

12. A doll is on sale for 9.9% off. If the original price is $12.99, what is the sale price?

13. The population of a city decreases by 42%. If the original population was 152,967, what would be the new population?

14. A cookie is now 92% fat free. If the regular cookie contains 6.2 grams of fat, how much fat does the "fat free" cookie have?

15. The number of pages in a magazine decreases by 78%. If the original magazine contained 99 pages, how many pages will the magazine now have?

Answers

1. $7.90

$$\begin{array}{r} \$19.75 \\ 1.99 \\ +\quad 0.36 \\ \hline \$22.10 \text{ Total purchases} \end{array}$$

$$\begin{array}{r} \$30.00 \\ -22.10 \\ \hline \$\ 7.90 \text{ change} \end{array}$$

2. $3.67

$$\begin{array}{r} \$2.99 \\ 1.99 \\ 0.99 \\ +\quad 0.36 \\ \hline \$6.33 \text{ total} \end{array} \qquad \begin{array}{r} \$10.00 \\ -\ 6.33 \\ \hline \$3.67 \text{ change} \end{array}$$

3. $160.60

6 arrangements × $39.90 = $239.40 total

$$\begin{array}{r} \$400.00 \\ -239.40 \\ \hline \$160.60 \text{ left} \end{array}$$

4. $10

$1999
2197
219            $4500
+ 75   − 4490
————   —————
$4490 total $ 10 left

5. $407

$ 599
179
52            $1500
+ 263   − 1093
————   —————
$1093 total $ 407 left

6. $1.34

New price = $1.25 + (0.07 × $1.25)
= $1.34

7. $15,106

New price = $13,000 + (0.162
× $13,000) = $15,106

8. 58,460 people

New attendance = 37,000 + (0.58
× 37,000) = 58,460 people

9. 141,840 people

New population = 47,123 + (2.01
× 47,123) = 141,840 people

10. 154 books

Number of books = 150 + (0.02572
× 150) = 154 books

11. $142.35

Sale price = $219 − (0.35 × $219)
= $142.35

12. $11.70

Sale price = $12.99 −

(0.099 × $12.99) = $11.70

13. 88,721 people

New population = 152,967 −
(0.42 × 152,967) = 88,721 people

14. 0.496 grams

Amount of fat = 6.2 − (0.92 × 6.2)
= 0.496 grams

15. 22 pages

Number of pages = 99 − (0.78 × 99)
= 22 pages

DATA ANALYSIS

Data analysis allows numerical values to be put into a picture form, such as bar graphs, line graphs, and circle graphs. In this manner, we gain a more intuitive understanding of the given information.

Bar graphs are used to compare amounts of the same measurements. In this example, we are comparing the number of bushels of corn produced on a farm from 1975–1985.

Example 1

Number of bushels (to the nearest 5 bushels) of wheat and corn produced by farm RQS from 1975–1985

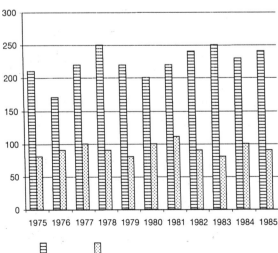

In which year was the least number of bushels of wheat produced? (See figure on preious page.)

By inspection of the graph, we find that the shortest bar representing wheat production is the one representing the wheat production for 1976. Thus, the least number of bushels of wheat was produced in 1976.

Line graphs are very useful in representing the factors of two different scales. Line graphs are often used to track the changes or shifts in certain factors. In the next example, the line graph is used to track the changes in the amount of scholarship money awarded to graduating seniors at a particular high school over the span of several years.

Example 2

Amount of scholarship money awarded to graduating seniors, West High, 1981–1990

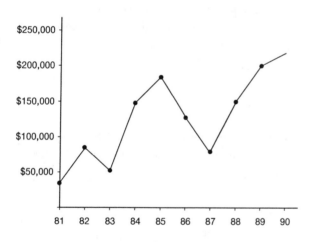

By how much did the scholarship money increase between 1987 and 1988?

To find the increase in scholarship money from 1987 to 1988, locate the amount for 1987 and 1988. In 1987 the amount of scholarship money is half-way between $50,000 and $100,000 or $75,000. In 1988 the

amount of scholarship money is $150,000. The increase is $150,000 − 75,000 = $75,000.

Circle graphs (or pie charts) are used to show the breakdown of a whole figure. When the circle graph is used to demonstrate a breakdown in terms of percents, the whole figure represents 100% and the parts of the circle graph represent pieces of the total that, when added together, add up to 100% (or the total). In this example, a circle graph shows how a family's budget has been divided into different categories using percents.

Example 3

Using the budget shown below, a family with an income of $1,500 a month would plan to spend what amount on housing?

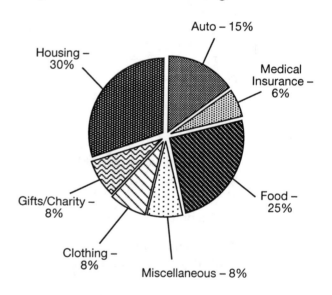

To find the amount spent on housing, locate on the pie chart the percentage allotted to housing, or 30%. 30% of $1,500 = $450.

Questions

Questions 1 and 2 refer to the following graph, which shows two students' scores on 5 math tests of 100 problems each.

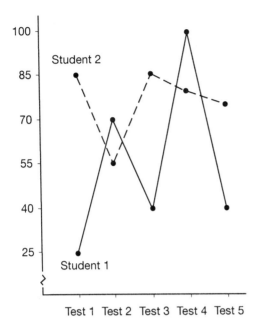

1. What is the difference between Student 2's Test 1 score and Student 1's Test 3 score?

2. What is the average of all five test scores for Student 1?

The following questions refer to the circle graph below, which shows the number of people born in the United States by age group.

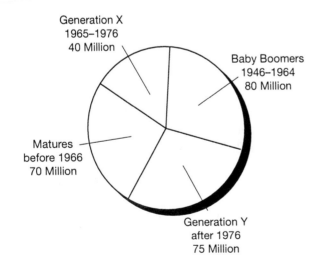

3. Approximately what percent of the entire population belongs to the Baby Boomers age group?

4. The combined population of the Matures age group and the Baby Boomers age group is what percent higher than the population of the Generation X age group?

A total of 72 people were surveyed and asked how many hours of sleep per night they require. The following questions refer to the bar graph below, which shows the results.

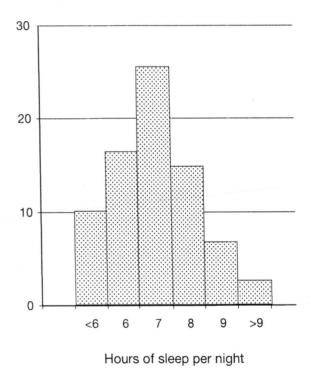

Hours of sleep per night

5. How many people needed more than 8 hours of sleep?

6. What is the ratio of the number of people who slept less than 7 hours per night to the number of people who slept more than 9 hours per night?

The following questions refer to the circle graph below, which shows how a family's monthly income of $4,000 is budgeted.

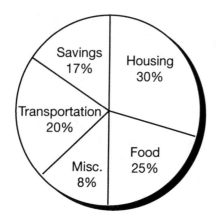

7. How much money is budgeted for savings and food combined for one month?

8. Over a 4-month period, what is the difference between the amount of money budgeted for housing and the amount of money budgeted for transportation?

For the following question, refer to the line graph on page 226.

9. What was the percent decrease in scholarship money awarded from 1986 to 1987?

The following question refers to the chart below.

Calorie Chart – Breads

Bread	Amount	Calories
French Bread	2 oz	140
Bran Bread	1 oz	95
Whole Wheat Bread	1 oz	115
Oatmeal Bread	0.5 oz	55
Raisin Bread	1 oz	125

10. If a person eats 3 oz. of bran bread and 2 oz. of raisin bread, what is the average number of calories consumed per ounce?

Answers

1. 45.

Student 2's Test 1 score is 85, whereas Student 1's Test 3 score is 40. 85 − 40 = 45.

2. 55.

Student 1's scores are: 25, 70, 40, 100, and 40.

(25 + 70 + 40 + 100 + 40) ÷ 5 = 275 ÷ 5 = 55.

3. 30%.

The total population for all age groups is (40 + 70 + 75 + 80) million = 265 million. The Baby Boomers represent 80 million. Then 80 million ÷ 265 million = 30%.

4. 275%.

The combined population of the Matures age group and the Baby Boomers age group is (70 + 80) million = 150 million. Since Generation X's population is 40 million, the difference is 110 million. Then 110 million ÷ 40 million = 275%.

5. 8 people.

There were 6 people who needed 9 hours of sleep and 2 people who needed more than 9 hours of sleep. Total of 8 people.

6. 13/1

There were 10 + 16 = 26 people who needed less than 7 hours of sleep, but only 2 people who needed more than 9 hours of sleep. Then, 26/2 becomes 13/1.

7. $1,680.

Savings and food combined is 17% + 25% = 42%. Then 42% of $4,000 equals $1,680.

8. $1,600.

Over a 4-month period, the cost for housing is (4) (0.30) ($4,000) = $4,800. During that same period of time, the cost for transportation is (4) (0.20) ($4,000) = $3,200. Then $4,800 – $3,200 = $1,600.

9. 40%.

$125,000 – $75,000 = $50,000

Then $50,000 ÷ $125,000 = 40%.

10. 107.

[(3) (95) + (2) (125)] ÷ 5 = 107.

☞ Practice: Advanced Topics

<u>**DIRECTIONS:**</u> Solve each problem below.

For numbers 1, 2, and 3, two ordinary six-sided dice are rolled.

1. The probability of rolling a 10 is _____.

2. The probability of rolling a number less than 5 is _____.

3. The probability of rolling a number greater than 8 is _____.

For numbers 4, 5, and 6: A jar contains two pennies, three nickels, five dimes, and six quarters. One coin is randomly selected from the jar.

4. The probability of selecting a nickel is _____.

5. The probability of selecting a penny or a dime is _____.

6. The probability of selecting a nickel or a quarter is _____.

7. In a study of people's work habits, 16 federal employees are chosen at random, and the number of days they worked during one month is determined. The average number of days worked per person is 20. If 12 of the people each worked 19 days during the month, how many days did each of the other four people work?

8. A class has four quizzes worth 30% (w1) of their grade, three tests worth 40% (w2), and a final worth 30% (w3). If these are Jenny's scores, what will her final grade average be, in percent?

quizzes: 67, 75, 85, and 72

tests: 85, 95, and 80

final: 83

9. If a triangle contains a 130° angle, it is called a(n) _____ triangle.

10. If one side of an equilateral triangle is 7", the perimeter = _____ inches.

11. In an isosceles triangle, one of the two equal sides is 5" and the third side is 3". The perimeter = _____ inches.

12. The circumference of a circle with a diameter of 5.4 inches = _____ inches (approx.).

13. The circumference of a circle with a radius of 3.5 feet = _____ feet (approx.).

14. The area of a circle with a radius of $1^2/_3$ inches = _____ inches^2 (approx.).

15. The area of a circle with a diameter of 9 inches = _____ inches^2 (approx.).

16. James buys a tennis racket for $40 and a can of tennis balls for $6.95. If he gives the clerk three $20 bills, the amount of change he gets back is _____.

17. Jean buys a pair of shoes for $43.95 and a pair of sneakers for $28.50. If the sales tax is $3.63 and she hands the clerk a $100 bill, her change should be _____.

18. The original price of a radio is $80. If the price increases by 12%, the new price will be _____.

19. A magazine costs $2.50. If the price increases by 160%, the new price will be _____.

20. A car that normally sells for $15,000 is being sold for 20% less. The reduced price will be _____.

21. A chair that normally sells for $60 is discounted by 32%. The discounted price will be _____.

22. The population of a city was 25,000 before a 16% decrease. The new population is _____.

For the following questions, refer to the graph below.

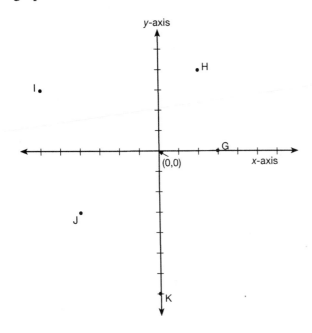

23. What are the coordinates of point K?

24. Starting at point I, move 2 units to the right, followed by 2 units down. How far will you be from point J?

For the following question, refer to the graph below.

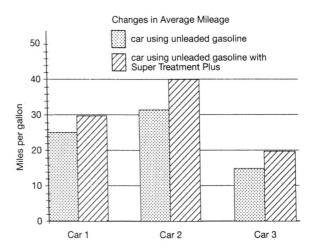

25. By how much did the mileage increase for Car 2 when the new product was used?

Answers

1. $\dfrac{1}{12}$

   The possible combinations are 6, 4; 4, 6; 5, 5. Then

   $$\frac{3}{16} = \frac{1}{12}.$$

2. $\dfrac{1}{6}$

   The six possible combinations are 1, 1; 1, 2; 2, 1; 1, 3; 3, 1; 2, 2. Then

   $$\frac{6}{36} = \frac{1}{6}.$$

3. $\dfrac{5}{18}$

   The ten ways are 4, 5; 5, 4; 4, 6; 6, 4; 5, 5; 3, 6; 6, 3; 5, 6; 6, 5; 6, 6. Then

$$\frac{10}{36} = \frac{5}{18}.$$

4. $\frac{3}{16}$

    Three nickels divided by 16 coins.

5. $\frac{7}{16}$

    Total of seven pennies and dimes divided by 16 coins.

6. $\frac{9}{16}$

    Total of nine nickels and quarters divided by 16 coins.

7. 23 days per person.

    (20) (16) = 320 man-days total. Then 320 − (12) (19) = 92 man-days left for 4 people. Finally, 92 ÷ 4 = 23 days per person.

8. 82%.

    (67 + 75 + 85 + 72) ÷ 4 = 74.75, which is the quiz average. (85 + 95 + 80) ÷ 3 = 86.67, which is the test average. Finally, (74.75) (0.30) + (86.67) (0.40) + (83) (0.30) = 81.993 = 82%, which becomes the final grade average.

9. obtuse

10. 21

    (7") (3) = 21"

11. 13

    (5") (2) + 3" = 13"

12. 16.956

    Circumference = (3.14) (5.4")
    = 16.956"

13. 21.98

    Circumference = (2) (3.14)
    (3.5 feet) = 21.98 feet

14. 8.72

    Area = (3.14) $(1^2/_3)^2$ = 8.72 inches^2

15. 63.585

    Area = (3.14) $(4.5)^2$
    = 63.585 inches^2

16. $13.05

    (3) ($20) − $40 − $6.95 = $13.05

17. $23.92

    $100 − $43.95 − $28.50 − $3.63
    = $23.92

18. $89.60

    ($80) (0.12) = $9.60. Then $80
    + $9.60 = $89.60

19. $6.50

    ($2.50) (1.60) = $4.00. Then $2.50
    + $4.00 = $6.50

20. $12,000

    ($15,000) (0.20) = $3,000. Then
    $15,000 − $3,000 = $12,000.

21. $40.80

    ($60) (0.32) = $19.20. Then $60
    − $19.20 = $40.80

22. 21,000

    (25,000) (0.16) = 4,000. Then
    25,000 − 4,000 = 21,000

23. (0, –7).

24. 4 units.

Point I's coordinates are (–6, 3). Moving 2 units to the right, then 2 units down will bring you to (–4, 1). Since the coordinates of point J are (–4, –3), the location of (–4, 1) is 4 units above point J. This is found by |– 3 – 1| or |1 – (–3)|.

25. 8 mpg.

The mileage for car 2 using the old product is 32 mpg from the bar graph. Using the new product the amount increases to 40 mpg. The amount of increase is 40 mpg – 32 mpg = 8 mpg.

REVIEW

Probability can be defined as the chance of something happening. This is often called the successful outcome. Probability of an event equals the ratio of successful outcomes to the total possible outcomes. This result must be between 0 and 1, inclusive. The probability of an impossible event is 0, and the probability of a certain event is 1.

In the basics of statistics presented here, it is most important to keep straight the definitions of mean, mode, and median. Remember that the arithmetic mean is essentially the average we learned earlier, the mode can be thought of as the "most popular" number, and the median is the "central" set of numbers.

In geometry, all triangles can be classified in one of several ways. Triangles can be defined as equilateral (whereby all three sides are equal), isosceles (having two sides equal), acute (all angles measure less than 90 degrees), obtuse (one angle measures more than 90 degrees), or right (having one right angle). A right angle measures 90 degrees. Two of these categories or classifications could apply simultaneously. For example, a geometry problem could feature an isosceles right triangle. The perimeter (or space around the outside) of a triangle is defined as the sum of all three sides.

In a circle, the diameter is defined as a line connecting two points on the circle and passing through the center of the circle. The radius is defined as a line going from the center of the circle to a point on the outside edge. The radius of a circle is equal to one-half of the circle's diameter. The distance around the outside of a circle is called its circumference. All circles have a set ratio between the circumference and the diameter. This ratio is expressed as the Greek letter pi (π). The value of pi is 3.14, or when written as a fraction, 22/7.

The circumference of circles equals $2\pi \times$ radius, or $\pi \times$ diameter.

Coordinate geometry allows us to map out figures based on points in space. By assigning the horizontal and vertical axes, we can plot several points and determine the distances between them. We can determine what quadrant a given point falls into, and we can also draw a line corresponding to a given algebraic equation.

Word problems can be solved by utilizing your mathematical skills in addition, subtraction, multiplication, division, geometry, percentages, and probability. Word problems often inject everyday, real-life situations into the world of mathematics. It is important to know how to utilize the mathematical concepts discussed in this book, not just to prepare you for the GED, but because in the "real" world, mathematics is a part of your everyday life.

Knowing how to read the various kinds of graphs presented in the Data Analysis section will prove invaluable as well. Knowing how to read a bar graph could help clarify a newspaper article; understanding a line graph could be helpful in investing money; and using a circle graph properly might be helpful in almost any work environment. From cooking to government spending, the ability to interpret data presented in a table or graph will prove as important everywhere else as on the GED.

Mathematics

Post-Test

MATHEMATICS

POST-TEST

DIRECTIONS: Choose the best answer choice for each question.

1. In the number 123,546,870, which digit is in the hundred thousands place?

   (1) 2

   (2) 3

   (3) 4

   (4) 5

   (5) 6

2. What is the number 643,581 rounded off to the thousands place?

   (1) 640,000

   (2) 643,000

   (3) 643,600

   (4) 644,000

   (5) 644,600

3. What is the number 0.0962 rounded off to the hundredths place?

   (1) 0.11

   (2) 0.10

   (3) 0.097

   (4) 0.096

   (5) 0.09

4. $10x^2 - 4x - 4x + x^2 = $ _____.

   (1) $11x^2$

   (2) $11x^2 - 8x$

   (3) $10x^4 - 8x$

   (4) $10x^4 - 4x^2$

   (5) $11x^2 + 8x$

5. What is the sum of

   $1\dfrac{3}{5}$ and $2\dfrac{2}{3}$?

   (1) $4\dfrac{2}{5}$

   (2) $4\dfrac{4}{15}$

   (3) $3\dfrac{5}{8}$

   (4) $3\dfrac{2}{5}$

   (5) $3\dfrac{4}{15}$

6. What is the value of

$$3\frac{1}{4} \div 4\frac{1}{6}?$$

(1) $\dfrac{3}{8}$

(2) $\dfrac{16}{27}$

(3) $\dfrac{39}{50}$

(4) $1\dfrac{11}{39}$

(5) 2

7. $(5xy)(7y - 6x) = \underline{\hspace{1cm}}$.

(1) $35xy^2 - 30x^2y$

(2) $5xy$

(3) $35xy - 30y$

(4) $5xy + 7y - 6x$

(5) $12xy^2 - 11x^2y$

8. $16m^{10} \div 8m^5 =$

(1) $2m^5$

(2) $2m^2$

(3) $8m^2$

(4) $2m^{15}$

(5) $8m^{15}$

9. What percent of 80 is 45?

(1) 25%

(2) 35%

(3) 56.25%

(4) 125%

(5) 177.78%

10. $0.9x + 3 > 0.6x$

(1) $x > -10$

(2) $x < 10$

(3) $x > 10$

(4) $x > -10$

(5) $x < -0.10$

11. Eighteen percent of some number is 24. What is the number?

(1) 0.0075

(2) 4.32

(3) 6

(4) 76

(5) $133\dfrac{1}{3}$

12. What is the square root of 0.0049?

(1) 7

(2) 4.9

(3) 0.7

(4) 0.49

(5) 0.07

13. A triangle in which exactly two sides are equal is described as \underline{\hspace{1cm}}.

(1) acute

(2) right

(3) equilateral

(4) obtuse

(5) isosceles

14. A triangle in which one of the angles is 132° is described as _____.

    (1) acute

    (2) right

    (3) equilateral

    (4) obtuse

    (5) isosceles

15. Given the numbers 6, 8, 10, and 20, what number when added would make the average of all five numbers 12?

    (1) 4

    (2) 10

    (3) 16

    (4) 22

    (5) 28

16. The perimeter of a triangle is 29. If two of the sides are 10 and 12, what is the length of the third side?

    (1) 7

    (2) 8

    (3) 9

    (4) 10

    (5) 11

17. A box of raisins used to cost $1.50. If the price increased by 14%, what is the new price?

    (1) $1.57

    (2) $1.60

    (3) $1.64

    (4) $1.68

    (5) $1.71

18. A car which originally sold for $16,000 is now being sold for 35% less. What will be the discounted price?

    (1) $12,500

    (2) $10,400

    (3) $8,800

    (4) $5,600

    (5) $4,500

19. The dimensions of a bathroom floor are 25 inches by 45 inches. What is the floor's area, in square inches?

    (1) 140

    (2) 625

    (3) 840

    (4) 1,125

    (5) 2,025

20. What is the volume of a box which measures 7 inches by 10 inches by 15 inches?

    (1) 1,050 inches^3

    (2) 656 inches^3

    (3) 157 inches^3

    (4) 64 inches^3

    (5) 32 inches^3

21. A can contains 18 marbles—six red, four white, three blue, and five black. One marble is drawn without looking. What is the probability of selecting a red or black marble?

    (1) $\dfrac{13}{18}$

    (2) $\dfrac{2}{3}$

(3) $\dfrac{11}{18}$

(4) $\dfrac{7}{18}$

(5) $\dfrac{1}{3}$

22. A pair of ordinary dice is rolled. What is the probability of rolling a total of 5?

    (1) $\dfrac{1}{9}$

    (2) $\dfrac{5}{36}$

    (3) $\dfrac{1}{6}$

    (4) $\dfrac{1}{3}$

    (5) $\dfrac{5}{6}$

23. What is the approximate area of a circle in which the diameter is 14 feet?

    (1) 44 feet^2

    (2) 88 feet^2

    (3) 132 feet^2

    (4) 154 feet^2

    (5) 616 feet^2

24. What is the number 783,459 rounded off to the thousands place?

    (1) 780,000

    (2) 783,000

    (3) 783,400

    (4) 783,500

    (5) 784,000

25. If $3x - 2 = \dfrac{1}{2}x + 2\dfrac{1}{2}$, then $x =$ _____.

    (1) $4\dfrac{1}{2}$

    (2) 5

    (3) $1\dfrac{4}{5}$

    (4) $3\dfrac{1}{2}$

    (5) 1

26. $(2m^2 - m - 1) \div (m - 1)$

    (1) $m + 1$

    (2) $2m + 1$

    (3) $2m - 1$

    (4) $m - 1$

    (5) $2m^2 + 1$

27. $10\dfrac{2}{3} - 3\dfrac{7}{8} =$

    (1) $7\dfrac{19}{24}$

    (2) $7\dfrac{5}{24}$

    (3) $6\dfrac{19}{24}$

    (4) $6\dfrac{5}{24}$

    (5) $6\dfrac{1}{24}$

28. What is the value of

$$\frac{1}{2} \div \frac{1}{3} \times \frac{1}{4}?$$

   (1) $\dfrac{1}{24}$

   (2) $\dfrac{1}{6}$

   (3) $\dfrac{3}{8}$

   (4) $\dfrac{8}{3}$

   (5) 6

29. What is the value of

$0.02 + 0.026 \div (0.2 \times 0.3 + 0.2)$?

   (1) 0.428

   (2) 0.35

   (3) 0.177

   (4) 0.12

   (5) 0.0575

30. $(x^2 + x - 30) \div (x - 5)$

   (1) $x - 6$

   (2) $x + 25$

   (3) $x + 6$

   (4) $x^2 - 6$

   (5) $x - 25$

31. If the numbers 9, 10, and 15 occur with the frequencies 6, 4, and 10, respectively, what is the mean?

   (1) 11.3

   (2) 12.2

   (3) 13.1

   (4) 8.9

   (5) 7.1

32. What percent of 28 is 0.7?

   (1) 0.025%

   (2) 2.5%

   (3) 4%

   (4) 25%

   (5) 40%

33. The total weight of 5 women and 3 men is 1100 pounds. If the mean weight of the women is 115 pounds, what is the mean weight, in pounds, of the men?

   (1) 131

   (2) 153

   (3) 175

   (4) 167

   (5) 129

34. 5.5% of some number is 1.98. What is the number?

   (1) 36

   (2) 18.89

   (3) 8

   (4) 3.8

   (5) 25.27

35. What is the cube root of 0.000027?

   (1) 0.0009

   (2) 0.003

   (3) 0.009

   (4) 0.03

   (5) 0.3

36. If $650 is placed into a savings account paying 8% interest for four years, how much simple interest is earned?

    (1) $52

    (2) $128

    (3) $208

    (4) $256

    (5) $280

37. If two angles of a triangle are 35° and 74°, what is the degree measure of the third angle?

    (1) 35°

    (2) 55°

    (3) 71°

    (4) 74°

    (5) 109°

38. Factor $x^2 - 6x + 9$.

    (1) $x(x + 6)$

    (2) $(x + 3)(x + 3)$

    (3) $(x - 3)^2$

    (4) $(x + 3)^2$

    (5) $(x + 3)(x - 3)$

39. The average of

    $\dfrac{1}{2}, \dfrac{1}{3}, \dfrac{1}{4}, \dfrac{2}{3}$, and $\dfrac{3}{8}$ is _____ .

    (1) $\dfrac{2}{5}$

    (2) $\dfrac{51}{120}$

(3) $\dfrac{1}{2}$

(4) $\dfrac{17}{24}$

(5) $\dfrac{11}{12}$

40. John used to weigh 140 pounds. If his weight increases by 15%, what is his new weight, in pounds?

    (1) 152

    (2) 155

    (3) 158

    (4) 161

    (5) 164

41. A pair of shoes was originally priced at $90. After two successive discounts of 10% and 5%, what was the final price?

    (1) $72.90

    (2) $75.00

    (3) $76.95

    (4) $78.25

    (5) $80.40

42. A rectangle has a length of 30 and a perimeter of 108. What is the width?

    (1) 78

    (2) 56

    (3) 48

    (4) 28

    (5) 24

43. A pair of ordinary dice is rolled. What is the probability that each die will show a number higher than 4?

   (1) $\dfrac{1}{36}$

   (2) $\dfrac{1}{12}$

   (3) $\dfrac{1}{6}$

   (4) $\dfrac{1}{4}$

   (5) $\dfrac{1}{3}$

44. A jar consists of 25 marbles of different colors. If the probability of selecting a blue marble is 0.28, how many blue marbles are in the jar?

   (1) 4

   (2) 5

   (3) 6

   (4) 7

   (5) 8

45. The circumference of a circle is 13. What is the approximate length of the diameter?

   (1) 2.3

   (2) 3.2

   (3) 4.1

   (4) 5.5

   (5) 6.4

46. John's test grades in his science class are 82, 93, 75, and 89. What is his average?

   (1) 87.45

   (2) 80.00

   (3) 90.40

   (4) 89.25

   (5) 84.75

47. In the number 985,027 what digit is in the thousands place?

   (1) 9

   (2) 2

   (3) 98

   (4) 8

   (5) 5

48. What is .047 written as a percent?

   (1) 4700%

   (2) 0.0047%

   (3) 470%

   (4) 4.7%

   (5) 47%

49. Which of the following could be the lengths of the sides of an isosceles triangle?

   (1) 8, 4, 8

   (2) 3, 2, 9

   (3) 8, 2, 9

   (4) 4, 5, 3

   (5) 3, 1, 7

50. In the pie graph below, what is the ratio of the amount spent by the category of all other departments to the combined amount spent by NASA, the Department of Defense, and the Department of Commerce?

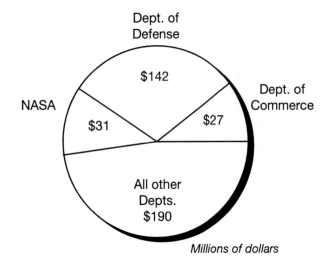

Millions of dollars

(1) $\dfrac{5}{11}$

(2) $\dfrac{8}{13}$

(3) $\dfrac{11}{15}$

(4) $\dfrac{14}{17}$

(5) $\dfrac{19}{20}$

For problems 51–53, choose the best method for solving each.

51. If $400 is placed into a savings account paying 6.5% interest for three years, how much simple interest is earned?

(1) $(400 \times 6.5) \times 3$

(2) $(400 \div 6.5) \times 3$

(3) $(400 \times 0.065) \times 3$

(4) $(400 + 0.65) \times 3$

(5) $(400 \times 0.65) \times 3$

52. What is the volume of a box which measures 6 feet by 14 feet by 20 feet?

(1) $6 + 14 + 20$

(2) $(6 + 14 + 20) \div 3$

(3) $6 \times 14 \times 20$

(4) $(6 + 14 + 20) \times (3)$

(5) $6 + 14 \times (20)$

53. A group of numbers consists of three 4's, four 5's, and seven 9's. What is the median?

(1) $(3 \times 4) + (4 \times 5) + (7 \times 9)$

(2) $(5 + 9) \div 2$

(3) $(5 - 9) \div 2$

(4) $(5 + 9) \times 2$

(5) $\dfrac{(3 \times 4) + (4 \times 5)}{(7 + 9)}$

54. What is the value of

$(20 - 3 \times 5) \div (6 + 5 \times 2)$?

55. What is 0.75% of 0.4?

56. If you start at point A on the graph below, and move 8 units in the negative *x* direction and 5 units in the negative *y* direction, how many units from point C would your next position be?

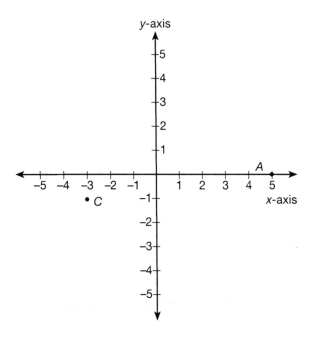

MATHEMATICS

ANSWER KEY

1. (4)	15. (3)	29. (4)	43. (2)
2. (4)	16. (1)	30. (3)	44. (4)
3. (2)	17. (5)	31. (2)	45. (3)
4. (2)	18. (2)	32. (2)	46. (5)
5. (2)	19. (4)	33. (3)	47. (5)
6. (3)	20. (1)	34. (1)	48. (4)
7. (1)	21. (3)	35. (4)	49. (1)
8. (1)	22. (1)	36. (3)	50. (5)
9. (3)	23. (4)	37. (3)	51. (3)
10. (1)	24. (2)	38. (3)	52. (3)
11. (5)	25. (3)	39. (2)	53. (2)
12. (5)	26. (2)	40. (4)	54. $\frac{5}{16}$
13. (5)	27. (3)	41. (3)	55. 0.003
14. (4)	28. (3)	42. (5)	56. 4 units

POST-TEST SELF-EVALUATION

Question Number	Subject Matter Tested	Section to Study (section, heading)
1.	Place Value	Whole Numbers, Introduction to Whole Numbers
2.	Rounding Numbers	Whole Numbers, Introduction to Whole Numbers
3.	Rounding Numbers	Whole Numbers, Introduction to Whole Numbers
4.	Algebra	Algebra, Operations with Polynomials—Addition, Subtraction
5.	Fractions	Fractions, Addition of Fractions
6.	Fractions	Fractions, Division of Fractions
7.	Algebra	Algebra, Operations with Polynomials—Multiplication
8.	Algebra	Algebra, Operations with Polynomials—Division
9.	Percents	Percents, Percent Problems—Working with Percents
10.	Algebra	Algebra, Inequalities
11.	Percents	Percents, Percent Problems—Working with Percents
12.	Square Roots	Topics in Mathematics, Cubes and Roots
13.	Triangles	Advanced Topics, Geometry
14.	Triangles	Advanced Topics, Geometry
15.	Averages	Topics in Mathematics, Averages
16.	Perimeter	Topics in Mathematics, Measurement
17.	Percents	Percents, Percent Problems—Working with Percents
18.	Percents	Percents, Percent Problems—Working with Percents
19.	Area	Topics in Mathematics, Measurement
20.	Volume	Topics in Mathematics, Measurement
21.	Probability	Advanced Topics, Probability
22.	Probability	Advanced Topics, Probability
23.	Circles	Advanced Topics, Geometry
24.	Rounding Numbers	Whole Numbers, Introduction to Whole Numbers
25.	Algebra	Algebra, Equations
26.	Algebra	Algebra, Operations with Polynomials—Division
27.	Fractions	Fractions, Subtraction of Fractions

Question Number	Subject Matter Tested	Section to Study (section, heading)
28.	Fractions	Fractions, Division of Fractions, Multiplication of Fractions
29.	Decimals	Decimals, Addition of Decimals
30.	Algebra	Algebra, Operations with Polynomials—Division
31.	Statistics	Advanced Topics, Statistics
32.	Percents	Percents, Percent Problems—Working with Percents
33.	Statistics	Advanced Topics, Statistics
34.	Percents	Percents, Percent Problems—Working with Percents
35.	Cube Roots	Topics in Mathematics, Cubes and Roots
36.	Percents	Percents, Percent Problems—Working with Percents
37.	Triangles	Advanced Topics, Geometry
38.	Algebra	Algebra, Simplifying Algebraic Expressions
39.	Averages	Topics in Mathematics, Averages
40.	Percents	Percents, Percent Problems—Working with Percents
41.	Percents	Percents, Percent Problems—Working with Percents
42.	Perimeter	Topics in Mathematics, Measurement
43.	Probability	Advanced Topics, Probability
44.	Probability	Advanced Topics, Probability
45.	Circles	Advanced Topics, Geometry
46.	Averages	Topics in Mathematics, Averages
47.	Place Value	Whole Numbers, Introduction to Whole Numbers
48.	Percents	Percents, Percent Problems—Working with Percents
49.	Triangles	Advanced Topics, Geometry
50.	Data Analysis	Advanced Topics, Data Analysis
51.	Percents	Percents, Percent Problems—Working with Percents
52.	Volume	Topics in Mathematics, Measurement
53.	Statistics	Advanced Topics, Statistics
54.	Order of Operations	Topics in Mathematics, Order of Operations
55.	Percents	Percents, Percent Problems—Working with Percents
56.	Graphs	Advanced Topics, Graphs

POST-TEST
ANSWERS AND EXPLANATIONS

1. **(4)** In a whole number, the hundred thousands place is the sixth digit from the right, which is 5.

2. **(4)** Since the digit in the hundreds place (5) is 5 or more, raise the thousands place digit (3) by 1, and change all digits to its right into zeros. This will yield 644,000.

3. **(2)** The digit 6, which is in the thousandths place, is 5 or more. This forces the digit 9 to increase by 1 (to a 0), and consequently change the tenths digit to a 1. The final answer is 0.10.

4. **(2)** Combine only like terms.

$10x^2 + x^2 = 11x^2$, and $-4x - 4x = -8x$, so the answer is $11x^2 - 8x$.

5. **(2)**

$$1\frac{3}{5} + 2\frac{3}{8} = \frac{8}{5} + \frac{8}{3} = \frac{24}{15} + \frac{40}{15} = \frac{64}{15} = 4\frac{4}{15}$$

6. **(3)**

$$3\frac{1}{4} \div 4\frac{1}{6} = \frac{13}{4} \div \frac{25}{6} = \frac{13}{4} \times \frac{6}{25} = \frac{78}{100} = \frac{39}{50}$$

7. **(1)** When multiplying, add the exponents

$$(5xy)(7y - 6x) = (5xy)(7y) + (5xy)(-6x)$$

The answer is $35xy^2 - 30x^2y$.

8. **(1)** When dividing, subtract exponents. $16 \div 8 = 2$, and $m^{10} \div m^5 = m^5$, so the answer is $2m^5$.

9. **(3)**

$$\frac{45}{80} = 0.5625 = 56.25\%$$

10. **(1)**

$$0.9x + 3 - 9x > 0.6x - 0.9x$$
$$3 \div -0.3x > -0.3x \div -0.3x$$
$$-10 < x$$

So $x > -10$.

11. **(5)**

$$\frac{24}{18\%} = \frac{24}{0.18} = 133\frac{1}{3}$$

12. **(5)**

0.07, since $0.07 \times 0.07 = 0.0049$.

13. **(5)** An isosceles triangle has two equal sides.

14. **(4)** An obtuse triangle contains one angle greater than 90°.

15. **(3)** If the average is 12, the sum of all five numbers must be

$$12 \times 5 = 60.$$

Then

$$60 - 6 - 8 - 10 - 20 = 16$$

16. **(1)**

$$29 - 10 - 12 = 7$$

17. **(5)**   The new price =

$$\$1.50 + \$1.50 \times 0.14 = \$1.71$$

18. **(2)**   The discounted price =

$$\$16,000 \times 1 - 0.35 = \$10,400$$

19. **(4)**   The area =

$$25 \times 45 = 1,125 \text{ square inches}$$

20. **(1)**   The volume =

$$7 \times 10 \times 15 = 1,050 \text{ inches}^3$$

21. **(3)**   Since there are 11 marbles which are either red or black, the probability is $^{11}/_{18}$.

22. **(1)**   The four ways of getting a total of 5 are: 1, 4; 4, 1; 2, 3; and 3, 2. Since there are $6 \times 6 = 36$ different outcomes, the probability is

$$\frac{4}{36} = \frac{1}{9}$$

23. **(4)**

$$\text{Area} = \pi \times (\text{radius})^2 = \pi \times 7^2$$

$$= 49\pi = 154 \text{ feet}^2$$

24. **(2)**   Since the hundreds digit 4 is less than 5, leave the thousands digit 3 as is. Change all digits to the right of 3 into zeros. This gives 783,000.

25. **(3)**

$$3x - 2 - \frac{1}{2}x + 2 = \frac{1}{2}x + 2\frac{1}{2} - \frac{1}{2}x + 2$$

$$2\frac{1}{2}x \div 2\frac{1}{2} = \frac{1}{2} \div 2\frac{1}{2}$$

$$x = \frac{9}{2} \div \frac{5}{2} = 1\frac{4}{5}$$

The answer is $x = 1\frac{4}{5}$.

26. **(2)**

$$
\begin{array}{r}
2m+1 \\
m-1 \overline{\smash{)}\ 2m^2 - m - 1} \\
\underline{-(2m^2 - 2m)} \\
m - 1 \\
\underline{-(m-1)} \\
0
\end{array}
$$

The answer is $2m + 1$.

27. **(3)**

$$10\frac{2}{3} - 3\frac{7}{8} = \frac{32}{3} - \frac{31}{8} = \frac{256}{24} - \frac{93}{24}$$

$$= \frac{163}{24} = 6\frac{19}{24}$$

28. **(3)**

$$\frac{1}{2} \div \frac{1}{3} \times \frac{1}{4} = \frac{1}{2} \times \frac{3}{1} \times \frac{1}{4} = \frac{3}{8}$$

29. **(4)**

$$0.2 \times 0.3 + 0.2 = 0.26$$

Then

$$0.02 + 0.026 \div 0.26 = 0.02 + 0.1 = 0.12$$

30. **(3)**

$$
\begin{array}{r}
x+6 \\
x-5 \overline{\smash{)}\ x^2 + x - 30} \\
\underline{-(x^2 - 5x)} \\
6x - 30 \\
\underline{-(6x - 30)} \\
0
\end{array}
$$

The answer is $x + 6$.

31. **(2)**

$$[(9)(6) + (10)(4) + (15)(10)] \div (6 + 4 + 10)$$

$$= 244 \div 20 = 12.2$$

32. **(2)**

$$\frac{0.7}{28} = 0.025 = 2.5\%$$

33. **(3)** The total weight of the women is (5) (115) = 575 pounds so the men's total weight must be 1,100 − 575 = 525 pounds. The average (mean) weight of the men is 525 ÷ 3 = 175 pounds.

34. **(1)**

$$1.98 \div 0.055 = 36$$

35. **(4)** The cube root is 0.03 because

$$(0.03)^3 = 0.000027$$

36. **(3)**

$$\$650 \times 0.08 \times 4 = \$208$$

37. **(3)** The third angle is

$$180° − 35° − 74° = 71°.$$

38. **(3)** This is the expression of the form $A^2 − 2AB + B^2$, which is the square of a binomial $(A − B)^2$. The answer is $(x − 3)^2$.

39. **(2)**

$$\frac{1}{2} + \frac{1}{3} + \frac{1}{4} + \frac{2}{3} + \frac{3}{8}$$

$$= \frac{12}{24} + \frac{8}{24} + \frac{6}{24} + \frac{16}{24} + \frac{9}{24} = \frac{51}{24}$$

Then

$$\frac{51}{24} \div 5 = \frac{51}{120}$$

40. **(4)**

$$140 \times 0.15 = 21$$

Then

$$140 + 21 = 161 \text{ pounds}$$

41. **(3)**

$$\$90 \times 0.10 = \$9;$$

$$\$90 − \$9 = \$81;$$

$$\$81 \times 0.05 = \$4.05$$

Finally,

$$\$81 − \$4.05 = \$76.95$$

42. **(5)**

$$30 \times 2 = 60; \quad 108 − 60 = 48;$$

then the width must be

$$48 \div 2 = 24.$$

43. **(2)** There are three successful outcomes:

5, 5; 5, 6; and 6, 5.

The probability is

$$\frac{3}{36} = \frac{1}{12}.$$

44. **(4)**

$$25 \times 0.28 = 7 \text{ marbles which are blue.}$$

45. **(3)** Since circumference = π × diameter, the diameter

$$= 13 \div π,$$

which is about 4.1.

46. **(5)** To determine averages, add all the numbers and divide by the number of numbers.

$$82 + 93 + 75 + 89 = 339$$

$$339 \div 4 = 84.75$$

47. **(5)** The thousands place is the fifth place from the decimal. In this case it is the number 5.

48. **(4)** To convert from decimal numbers to a percent, move the decimal point two places to the right.

49. **(1)** An isosceles triangle has two sides of equal length.

50. **(5)** The total for all other departments is $190. The combined total for NASA, the Department of Defense, and the Department of Commerce is $200.

So, $\dfrac{190}{200} = \dfrac{19}{20}$

51. **(3)** In order to determine the interest, use the formula:

Interest = Principle × Rate × Time

$400 × 0.065 × 3

52. **(3)** In order to determine volume of a box, use the following formula:

Volume = length × width × height

6 × 14 × 20

53. **(2)** Since there are a total of 14 numbers (3 + 4 + 7), the median is the average of the seventh and eight numbers. Arranged in order, the seventh number is 5 and the eighth number is 9. Thus, the solution for the median is (5 + 9) ÷ 2

54. Follow the order of operations to get

$(20 - 3 \times 5) \div (6 + 5 \times 2)$

$(20 - 15) \div (6 + 10) = \dfrac{5}{16}$

55. When working with problems involving words, "of" normally means multiplication. Also, the percents must be converted into decimals. The problem should be set up as follows.

0.75% = 0.0075

Then

0.0075 × 0.4 = 0.003

56. Point A is located at (5, 0). Moving 8 units in the negative x direction and 5 units in the negative y direction, your new location is at (–3, –5). Since point C is located (–3, –1), the required distance is $\lvert -1 - (-5) \rvert = 4$.

Mathematics

Glossary

MATHEMATICS

GLOSSARY OF SYMBOLS, TERMS, AND FORMULAS

Symbols

= —equal to. Symbol showing that two expressions are the same. Example: 2 + 3 = 5.

> —greater than. Symbol showing that one expression is greater than another. The symbol always "points" to the smaller expression. Example: 2 + 3 > 4.

< —less than. Symbol showing that one expression is less than another. The symbol always "points" to the smaller expression. Example: 2 + 3 < 6.

. —decimal point. Used to separate the whole number part from the decimal part of a decimal number.

— or **/** —fraction bar. Used to separate the numerator and denominator of a fraction.

( ) —parentheses. Used as grouping symbols in a mathematical expression.

{ } —braces. Used as grouping symbols in a mathematical expression.

[ ] —brackets. Used as grouping symbols in a mathematical expression.

% —percent.

| | —absolute value.

ø —empty set.

Terms

absolute inequality —one whose validity is true regardless of the value of the variables (within the set of reals) in the sentence.

absolute value —a given number, regardless of sign. Example: the absolute value of –3 is 3; the absolute value of 3 is also 3.

account —record of transactions involving money.

acute angle —angle measuring less than 90 degrees.

acute triangle —triangle containing three acute angles.

addition —operation of combining two numbers to get a third number called the sum.

algorithm —series of steps for calculating an answer.

amount —number resulting from a calculation.

angle —figure formed by two lines starting at the same point. Example:

answer line —answer calculated as part of the final answer.

approximate—to estimate; to make a best guess. Example: We can approximate 0.3 with $^1/_3$.

area—region inside a figure.

average—found by adding a group of numbers together and dividing this sum by the quantity of numbers. Example: The average of 2, 6, and 7 is

$$\frac{2+6+7}{3} = 5.$$

base—number being raised to an exponent. Example:

2 is the base in 2^4.

beginning (amount)—amount before a percentage is taken of it.

bimodal—set of numbers that has two or more modes.

binomial—algebraic expression with two terms.

bond account—account containing legal promises to repay the account holder's money.

borrow—in subtraction, the process of taking ten from the next column of digits to the left. Example:

$$\begin{array}{r} 27 \\ -18 \\ \hline \end{array}$$

Since 8 cannot be subtracted from 7, 10 is borrowed from the tens column. So, 8 can be subtracted from 17. Therefore,

$$\begin{array}{r} 27 \\ -18 \\ \hline 9 \end{array}.$$

box—three-dimensional figure containing all 90-degree angles.

braces—grouping symbols, { }, used in a mathematical expression.

brackets—grouping symbols, [ ], used in a mathematical expression.

calculate—compute, figure, determine.

cancellation—in an expression using addition or subtraction, removal of the same number with opposite signs. Example:

$$6 + 2 - 3 - 2 = 3$$

In an expression using division, removal of the same number that appears in the numerator and denominator. See "reduce." Example:

$$\frac{8}{12} = \frac{\cancel{4} \times 2}{\cancel{4} \times 3} = \frac{2}{3}.$$

carry—in addition and multiplication, the process of adding a number to the next column of digits to the left. Example:

$$\begin{array}{r} \overset{1}{} \\ 27 \\ +16 \\ \hline 43 \end{array}$$

Since $7 + 6 = 13$, 1 is carried to the tens place column.

centimeter—one hundredth of a meter. A measurement in the metric system.

certificate of deposit (CD)—document showing ownership of deposited money.

checking account—account against which checks are written.

circle—figure in which every point is the same distance from the center of the figure. Example:

circumference—measurement of the distance around a circle. See "perimeter."

coefficient—the constant part of a term that contains both constants and variables; 1 when a term has no constants.

columns format—aligning two or more numbers in columns so that their place values appear in the same column.

conditional inequality—one whose validity depends on the value of the variables in the sentence.

constant—a symbol which takes on only one value at a time.

consumer mathematics—area of mathematics dealing with purchases, savings, and other consumer-related transactions.

convert—change to another unit of measure. For example, you can convert 1 inch to 2.54 centimeters.

coordinate axes—pair of perpendicular number lines on a two-dimensional Cartesian coordinate plain.

coordinate geometry—study of geometric figures using algebraic principles.

cube—third power of a number. Example: 8 is the cube of 2.

cube root—number that when cubed results in the original number. Example: 2 is the cube root of 8. That is,

$2^3 = 8.$

cubic unit—unit used to measure a three-dimensional object such as a box. Example: a box might have a volume of 100 cubic feet.

decimal number—number made up of a whole number part and a decimal part. Example: In 6.42, 6 is the whole number part and 42 is the decimal part.

decimal part—that part of a number to the right of the decimal point. Example: 893 is the decimal part of the number 640.893. See "whole number part."

decimal place—position of a digit in the decimal part of a number. Position to the right of the decimal place. Example: using the decimal point name 0.8519:

first	tenths	8
second	hundredths	5
third	thousandths	1
fourth	ten thousandths	9

decimal point—symbol (a period) separating the whole number part from the decimal part of a number. Example: In 640.893, the decimal point separates 640 (whole part) from 893 (decimal part).

degree—unit of measure for a circle or an angle. There are 360 degrees in a circle.

denominator—that part of a fraction below the fraction bar. The bottom part of a fraction. See "numerator." Example: 3 is the denominator in $2/3$.

deposit—amount of money placed in an account.

diameter—in a circle, a line touching two points on the circle and the center of the circle. Example:

digit—any of the characters 0, 1, 2, 3, 4, 5, 6, 7, 8, 9.

dimension—measurement of a figure. Example: We can measure the length and width of a rectangle and the diameter of a circle.

dividend—number being divided. Example: 20 is the dividend in 20 ÷ 6.

divisible—able to be divided.

division—operation of finding how many times a number is found in another number. Example: 63 divided by 7 is 9. That is, there are nine 7's in 63.

divisor—dividing number. Example: 5 is the divisor in 26 ÷ 5.

empty set—equation without a solution.

equal symbol— =, in an equation showing that two or more expressions are exactly the same. Example: Notice the use of the equal symbol in 2 + 3 = 1 + 4 = 5.

equation—mathematical statement showing two or more equal expressions. Example: 2 + 3 = 1 + 4 = 5 is an equation.

equilateral triangle—triangle in which all sides are equal and all angles are equal.

equivalent equations—those that share the same solutions.

equivalent inequalities—those that share the same solutions.

even integers—set of integers divisible by two.

event—in probability, an occurrence of a certain action or happening. Example: When flipping a coin, the coin landing heads up is an event.

exponent—number appearing to the upper right of another number called the base. The exponent represents the number of times the base appears as it is multiplied by itself. Example: In 4^3, 3 is the exponent.

expression—combination of numbers and operations; in algebra, a collection of one or more terms. Example:

$$\frac{683 - 252}{59} \times 2$$

factor—number that divides into another number. Example: 6 is a factor of 42.

factor tree—picture of prime factors of a number. Example: The prime factors of 12 are 2 and 3, as determined below.

flipping—in a fraction, taking the reciprocal, that is, exchanging the numerator and denominator. Example: Flipping $^2/_3$ results in $^3/_2$. See "reciprocal."

formula—rule. Equation that is used to calculate some measurement of a figure. See Formulas section.

fraction—expression showing one number divided by the other. The numerator (top number) is separated from the denominator (bottom number) by the fraction bar. Example: $^2/_3$.

fraction bar—symbol, /, in a fraction separating the numerator and the denominator.

fractional part—in a mixed number, the fraction appearing after the whole number. Example: In $1\frac{2}{3}$, $\frac{2}{3}$ is the fractional part.

geometry—the study of measurements and properties of figures.

greater than—symbol, >, showing one expression larger than another. Example: 6 + 4 > 8.

greatest common factor (GCF)—largest factor common to all members of a given set.

group—in a mathematical expression, numbers and operations enclosed in grouping symbols. A group is simplified before the rest of the expression.

grouping symbol—symbols used with the Order of Operations to simplify an expression. See "braces," "brackets," and "parentheses."

hundred millions place—in a decimal number, the ninth digit to the left of the decimal point.

hundred thousands place—in a decimal number, the sixth digit to the left of the decimal point.

hundreds place—in a decimal number, the third digit to the left of the decimal point.

hundredths place—in a decimal number, the second digit to the right of the decimal point.

improper fraction—fraction in which the numerator is greater than the denominator.

index—number at the upper left of a root symbol. Example: 3 is the index in

$$\sqrt[3]{8}.$$

integers—set made up of the positive and negative whole numbers.

interest (rate)—percentage paid for the use of money.

interior angle—angle inside a figure.

isosceles triangle—triangle in which at least two sides are equal and at least two angles are equal.

least common multiple (LCM)—smallest quantity divisible by every number of a given set.

less than symbol—<, showing one expression smaller than another. Example:

6 + 2 < 10.

long division format—arrangement of numbers used in calculating a quotient. Example: $17\overline{)51}$ is equivalent to $\frac{51}{17} = 3$.

lowest common denominator (LCD)—smallest number into which denominators of two or more fractions will divide. Example:

LCD for $\frac{4}{5}$ and $\frac{1}{3}$ is 15.

median—middle number in a sequence.

meter—basic measurement in the metric system.

millions place—in a decimal number, the seventh digit to the left of the decimal point.

mixed number—number made up of a whole number part and a fractional part. Example: $2\frac{1}{4}$.

mode—number that appears most frequently in a set of numbers.

monomial—algebraic expression with only one term.

multiplication—operation of repeatedly adding a number to itself. Example: 2 multiplied by 3 means add three 2's together: $2 \times 3 = 2 + 2 + 2 = 6$.

multi-step word problem—word problem in which the solution is found by breaking down the problem into smaller steps.

mutual fund—account that invests in a variety of stocks.

natural numbers—set of integers starting with 1 and increasing.

negative numbers—set of integers starting with –1 and decreasing.

null set—see "empty set."

numerator—that part of a number above the fraction bar. The top part of a fraction. See "denominator." Example: 2 is the numerator in $\frac{2}{9}$.

obtuse angle—angle measuring more than 90 degrees.

obtuse triangle—triangle which contains an obtuse angle.

odd integers—set of integers not divisible by two.

of—in mathematics, "% of" means multiply. Example:

50% of 60 is 30.

ones place—in a decimal number, the first digit to the left of the decimal point.

operations—addition, subtraction, multiplication, division, and exponentiation.

order of operations—in a mathematical expression, order in which operations are performed.

Order—Operations

1st—Parentheses

2nd—Exponents

3rd—Multiplication and division

4th—Addition and subtraction

Apply the Order of Operations to any groups first.

origin—point of intersection of the x- and y-axes; see coordinate axes.

outcome—in probability, the result of an experiment. Example: The experiment is flipping a coin. There are two outcomes: the coin lands heads up; the coin lands tails up.

parentheses—grouping symbols, ( ), used in a mathematical expression.

percent—rate applied to a number. The percent (%) is a number between 0 and 100. Example: 50% of 40 is 20.

perimeter—sum of the lengths of the sides of a figure. See "area."

pi—Greek letter that denotes a number with a value of approximately 3.14159265.

place holder—in multiplication, 0 placed at the right end of an answer line.

place value—digit's position in a number. Example: In the number 748.1239, 8 is the value in the ones place, 1 is the value in the tenths place, and 2 is the value in the hundredths place.

polynomial—algebraic expression with two or more terms.

positive numbers—see "natural numbers."

prime factor—factor of a number that is also a prime number. Example: Prime factors of 12 are 2 and 3.

prime number—whole number that can only be divided evenly by two numbers: itself and 1. Prime numbers are 2, 3, 5, 7, 11 and so on. *Note:* 1 is *not* a prime number.

principle—beginning amount to be invested.

probability—event's chance of happening expressed as a number between 0 and 1, usually as a percent. Example: There is a 50% chance of rain tomorrow.

process—series of steps to complete a calculation.

proper fraction—fraction in which the numerator is less than the denominator.

quotient—result of a division. Example: 4 is the quotient of 20 divided by 5.

radius—in a circle, a line touching one point on the circle and the center of the circle. Example:

rate—percent.

ratio—division of two numbers. Example: A ratio can be written as $2:3 = \dfrac{2}{3}$.

real numbers (reals)—set of numbers found on the real number line, which stretches forever in each direction.

reciprocal—flipping a fraction, that is, exchanging the numerator and denominator. Example: The reciprocal of $\dfrac{3}{4}$ is $\dfrac{4}{3}$.

rectangle—four-sided figure with all angles measuring 90 degrees. Example:

reduce—in division, removal of the same number that appears in the numerator and denominator. See "cancellation." Example:

$$\frac{8}{12} = \frac{4 \times 2}{4 \times 3} = \frac{2}{3}.$$

remainder (R)—part of the dividend that remains because it is less than the divisor. Example:

$$\frac{23}{5} = 4 \text{ R3}$$

3 is the remainder since it is less than 5.

repeating decimal—decimal number in which one or more digits repeat infinitely. A bar appears above the repeating digit(s). Example: $\overline{0.36}$ means 0.363636...

right angle—angle formed by two lines forming a 90 degree angle. Example:

right triangle—triangle containing a right angle.

root—see "cube root" and "square root."

rounding—approximating a number by this process: If the digit to be rounded is greater than or equal to 5, drop this digit and increase the next digit to the left by 1. If the digit to be rounded is less than 5, drop that digit. Example:

5.8528 rounds to 5.853

7.8342 rounds to 7.834

sense—characteristic shared by inequalities whose signs of inequality point in the same direction.

sets—group of numbers useful in categorization.

simplify—use the Order of Operations to rewrite an expression.

solution—in an equation containing a single variable, a number that makes the equation true when substituted for the variable.

square—four-sided figure with all angles measuring 90 degrees, and all sides are equal. Example:

square root—number which when squared results in the original number. Example: 2 is the square root of 4: That is, $\sqrt{4} = 2$.

square unit—unit used to measure a two-dimensional object such as a rectangle. Example: A rectangle might have an area of 100 square feet.

stock—money raised by a company through the sale of shares.

subset—a group within a set.

subtraction—operation of finding the difference between two numbers.

successful outcome—outcome whose probability you want to find.

sum—total. Result of addition.

ten millions place—in a decimal number, the eighth digit to the left of the decimal point.

ten thousands place—in a decimal number, the fifth digit to the left of the decimal point.

ten thousandths place—in a decimal number, the fourth digit to the right of the decimal point.

tens place—in a decimal number, the second digit to the left of the decimal point.

tenths place—in a decimal number, the first digit to the right of the decimal point.

term—a constant, variable, or combination of both. Examples: 3*x*, 4, and *yz* are all terms.

test fraction—fraction used to determine the Least Common Denominator.

thousands place—in a decimal number, the fourth digit to the left of the decimal point.

thousandths place—in a decimal number, the third digit to the right of the decimal point.

time—period for which an investment is made.

triangle—three-sided figure. The sum of the internal angles of a triangle is 180 degrees.

trinomial—algebraic expression with three terms.

unit—standard of measurement. Example: inch, meter.

value—see "place value."

variable—a placeholder which can take on any of several values at a given time.

volume—region inside a three-dimensional figure like a box. See "area."

whole number part—in a mixed number, the whole number appearing before the fraction. Example: In 6.13, 6 is the whole number part. In a decimal number, the whole number to the left of the decimal point. Example: In 932.17, 932 is the whole number part.

whole numbers—natural numbers with 0. That is: 0, 1, 2, 3, and so on.

withdraw—take money from an account.

x-axis—horizontal axis; see coordinate axes.

y-axis—vertical axis; see coordinate axes.

Formulas

Amount = percent × beginning (amount)—Final amount after taking a percent of the beginning amount

Area = length × width—Area of a rectangle

Area = side squared = Area of a square

Area = pi × radius squared—Area of a circle

Circumference = pi × diameter—Circumference (perimeter) of a circle

Interest = principal × rate × time—Simple interest

Perimeter = 2 × length + 2 × width—Perimeter of a rectangle

Perimeter = 4 × side—Perimeter of a square

Probability = successful outcome ÷ total possible outcomes—Probability of a successful outcome

Mathematics

Appendix

MATHEMATICS REFERENCE TABLE

SYMBOLS AND THEIR MEANINGS

=	is equal to	$\leq$	is less than or equal to
$\neq$	is unequal to	$\geq$	is greater than or equal to
<	is less than	||	is parallel to
>	is greater than	$\perp$	is perpendicular to

FORMULAS

DESCRIPTION	FORMULA
AREA (*A*) of a:	
square	$A = s^2$; where s = side
rectangle	$A = lw$; where l = length, w = width
parallelogram	$A = bh$; where b = base, h = height
triangle	$A = \frac{1}{2}bh$; where b = base, h = height
circle	$A = \pi r^2$; where $\pi = 3.14$, r = radius
PERIMETER (*P*) of a:	
square	$P = 4s$; where s = side
rectangle	$P = 2l + 2w$; where l = length, w = width
triangle	$P = a + b + c$; where a, b, and c are the sides
circumference (*C*) of a circle	$C = \pi d$, where $\pi = 3.14$, d = diameter
VOLUME (*V*) of a:	
cube	$V = s^2$; where s = side
rectangular container	$V = lwh$; where l = length, w = width, h = height
Pythagorean relationship	$c^2 = a^2 + b^2$; where c = hypotenuse, a and b are legs of a right triangle
distance (*d*) between two points in a plane	$d = \sqrt{(x_2 - x_1)^2 + (y_2 - y_1)^2}$ where (x_1, y_1) and (x_2, y_2) are two points in a plane
mean	**mean** $= \dfrac{x_1 + x_2 + ... + x_n}{n}$; where the x's are the values for which a mean is desired, and n = number of values in the series
median	**median** = the point in an ordered set of numbers at which half of the numbers are above and half of the numbers are below this value
simple interest (*i*)	$i = prt$; where p = principal, r = rate, t = time
distance (*d*) as function of rate and time	$d = rt$; where r = rate, t = time
total cost (*c*)	$c = nr$; where n = number of units, r = cost per unit

REQUIREMENTS FOR ISSUANCE OF CERTIFICATE/DIPLOMA

Location	Minimum Test Score	Minimum Age For Credential	Residency Requirement	Minimum Age For Testing	Testing Fee Per Battery	Title Of Credential
UNITED STATES						
Alabama	40 min & 45 avg	18[1]	30 days	18[1]	$25.00[2]	Cert. of H.S. Equiv.
Alaska	40 min & 45 avg	18[1]	resident	18[1]	max. $25.00	H.S. Dipl.
Arizona	40 min & 45 avg	18[1]	none	18[1]	max. $25.00[2]	H.S. Cert. of Equiv.
Arkansas	40 min & 45 avg	16	legal resident	16[1]	none	H.S. Dipl.
California	40 min & 45 avg	18[1]	resident	18[1]	varies	H.S. Equiv. Cert.
Colorado	40 min & 45 avg	17	resident[1]	17	$25.00-$40.00	H.S. Equiv. Cert.
Connecticut	40 min & 45 avg	17[1]	resident	17[1]	over 21, $13.00	H.S. Dipl.
Delaware	40 min & 45 avg	18	resident	18[1]	$25.00	St. Bd. of Ed. Endsmt.
District of Columbia	40 min & 45 avg	18	resident[1]	18[1]	$20.00	H.S. Equiv. Cert.
Florida	40 min & 45 avg	18	legal resident	18[1]	$25.00	H.S. Dipl.
Georgia	40 min & 45 avg	18[1]	none	18[1]	$35.00	Gen. Educ. Dev. Diploma

[1]Jurisdictional requirements on exceptions and limitations.
[2]Jurisdictional requirements on credential-related and other fees.

Location	Minimum Test Score	Minimum Age For Credential	Residency Requirement	Minimum Age For Testing	Testing Fee Per Battery	Title Of Credential
Hawaii	40 min & 45 avg	17	resident[1]	17[1]	$20.00	Dept. of Ed. H.S. Dipl.
Idaho	40 min & 45 avg	18	resident	18[1]	varies	H.S. Equiv. Cert.
Illinois	40 min & 45 avg	18[1]	30 days	18[1]	$15.00[2]	H.S. Equiv. Cert.
Indiana	40 min & 45 avg	17[1]	30 days	17[1]	maximum $25.00	H.S. Dipl.
Iowa	40 min & 45 avg	17[1]	none	17[1]	$20.00[2]	H.S. Equiv. Dipl.
Kansas	40 min & 45 avg	16[1]	resident[1]	16[1]	$30.00	H. S. Dipl.
Kentucky	40 min & 45 avg	16	resident	16[1]	$25.00	H.S. Equiv. Dipl.
Louisiana	40 min & 45 avg	17[1]	resident[1]	17[1]	maximum $20.00	H.S. Equiv. Dipl.[1]
Maine	40 min & 45 avg	18[1]	none	18[1]	none[2]	H.S. Equiv. Dipl.
Maryland	40 min & 45 avg	16[1]	3 months	16	$18.00[2]	H.S. Dipl.
Massachusetts	40 min & 45 avg	19[1]	resident	19[1]	$40.00	H.S. Equiv. Cert.
Michigan	40 min & 45 avg	18[1]	30 days	16[1]	varies	H.S. Equiv. Cert.
Minnesota	40 min & 45 avg	19[1]	resident	19[1]	$40.00	Sec. Sch. Equiv. Cert.

[1]Jurisdictional requirements on exceptions and limitations.
[2]Jurisdictional requirements on credential-related and other fees.

Location	Minimum Test Score	Minimum Age For Credential	Residency Requirement	Minimum Age For Testing	Testing Fee Per Battery	Title Of Credential
Mississippi	40 min & 45 avg	17	30 days[1]	17[1]	$20.00	H.S. Equiv. Dipl.
Missouri	40 min & 45 avg	16[1]	resident[1]	16[1]	$20.00	Cert. of H.S. Equiv.
Montana	40 min & 45 avg	17[1]	resident[1]	17[1]	$18.00	H.S. Equiv. Cert.
Nebraska	40 min & 45 avg	18	30 days[1]	16[1]	$20.00–$30.00[2]	Dept. of Ed. H.S. Dipl.
Nevada	40 min & 45 avg	17	none	17	$25.00[2]	Cert. of H.S. Equiv.
New Hampshire	40 min & 45 avg	18	resident	18[1]	$40.00	Cert. of H.S. Equiv.
New Jersey	40 min & 45 avg	16[1]	resident	16[1]	$20.00	H.S. Dipl.
New Mexico	40 min & 45 avg	18[1]	resident	18[1]	varies[2]	H.S. Dipl.
New York	40 min & 45 avg	19[1]	1 month	19[1]	none	H.S. Equiv. Dipl.
North Carolina	40 min & 45 avg	16	resident[1]	16[1]	$7.50[2]	H.S. Dipl. Equiv.
North Dakota	40 min & 45 avg	18[1]	none	18[1]	varies	H.S. Equiv. Cert.
Ohio	40 min & 45 avg	19[1]	resident	19	$42.00[1,2]	Cert. of H.S. Equiv.

[1]Jurisdictional requirements on exceptions and limitations.
[2]Jurisdictional requirements on credential-related and other fees.

Location	Minimum Test Score	Minimum Age For Credential	Residency Requirement	Minimum Age For Testing	Testing Fee Per Battery	Title Of Credential
Oklahoma	40 min & 45 avg	16[1]	resident	16[1]	varies[2]	Cert. of H.S. Equiv.
Oregon	40 min & 45 avg	18[1]	resident[1]	18[1]	varies[2]	Cert. of Gen. Ed. Dev.
Pennsylvania	40 min & 45 avg[1]	18[1]	resident[1]	18[1]	varies	Com. Sec. Sc. Dipl.
Rhode Island	40 min & 45 avg	16[1]	resident	16[1]	$15.00	H.S. Equiv. Dipl.
South Carolina	40 min & 45 avg	17	resident[1]	17[1]	varies	H.S. Equiv. Dipl.
South Dakota	40 min & 45 avg	18[1]	resident[1]	17[1]	maximum $20.00	H.S. Equiv. Cert.
Tennessee	40 min & 45 avg	18[1]	resident	18[1]	$20.00-$25.00	Equiv. H.S. Dipl.
Texas	40 min & 45 avg	18[1]	resident[1]	18[1]	varies[2]	Cert. of H.S. Equiv.
Utah	40 min & 45 avg	17[1]	resident[1]	17[1]	$25.00 and up	Cert. of Gen. Ed. Dev.
Vermont	40 min & 45 avg	16	none	16[1]	$25.00-$30.00	Sec. Sc. Equiv. Cert.
Virginia	40 min & 45 avg	18[1]	resident	18[1]	$25.00[2]	Com. Gen. Ed. Dev. Cert.
Washington	40 min & 45 avg	19[1]	resident	19[1]	$25.00	Cert. of Ed. Comp.
West Virginia	40 min & 45 avg	18[1]	30 days	18[1]	varies	H.S. Equiv. Dipl.

[1]Jurisdictional requirements on exceptions and limitations.
[2]Jurisdictional requirements on credential-related and other fees.

Location	Minimum Test Score	Minimum Age For Credential	Residency Requirement	Minimum Age For Testing	Testing Fee Per Battery	Title Of Credential
Wisconsin	40 min & 45 avg	18	voting resident	18[1]	varies	H.S. Equiv. Dipl.
Wyoming	40 min & 45 avg	18	resident[1]	17[1]	varies	H.S. Equiv. Cert.
CANADA - PROVINCES & TERRITORIES						
Alberta	45 min each test	18[1]	resident	18	$50.00	H.S. Equiv. Dipl.
British Columbia	45 min each test	19[1]	resident	19	$26.75	Sec. Sc. Equiv. Cert.
Manitoba	45 min each test	19[1]	resident	19	$22.00	H.S. Equiv. Dipl.
New Brunswick	45 min each test	19	resident	19	$10.00	H.S. Equiv. Cert.
Newfoundland	45 min each test	19[1]	resident	19	none	H.S. Equiv. Dipl.
Northwest Terr.	45 min each test	18[1]	6 months	18[1]	$5.00	H.S. Equiv. Cert.
Nova Scotia	45 min each test	19[1]	none	19	$20.00	H.S. Equiv. Dipl.
Prince Edward Is.	45 min each test	19[1]	resident	19[1]	$20.00	H.S. Equiv. Cert.
Saskatchewan	45 min each test[1]	19	resident	19[1]	$25.00	H.S. Equiv. Cert.

[1]Jurisdictional requirements on exceptions and limitations.
[2]Jurisdictional requirements on credential-related and other fees.

Location	Minimum Test Score	Minimum Age For Credential	Residency Requirement	Minimum Age For Testing	Testing Fee Per Battery	Title Of Credential
Yukon	45 min each test	19[1]	resident	19[1]	$25.00	Sec. Sc. Equiv. Cert.
U.S. TERRITORIES						
American Samoa	40 min & 45 avg	17	resident	17[1]	$20.00	H.S. Dipl. of Equiv.
Guam	40 min & 45 avg	18	resident	18[1]	$10.00	H.S. Equiv. Dipl.
Kwajalein	40 min & 45 avg	18	resident	18	$27.50	H. S. Equiv. Dipl.
Mariana Islands	40 min & 45 avg	18[1]	30 days	18[1]	$5.00[2]	H.S. Equiv. Dipl.
Marshall Islands	40 min & 45 avg	17[1]	30 days	17	$7.50[2]	H.S. Equiv. Dipl.
Micronesia	40 min & 45 avg	18	resident	18[1]	$7.50[2]	H.S. Equiv. Cert.
Palau	40 min & 45 avg	16	Contact your local Dept. of Ed.	16[1]	$10.00	Cert. of Equiv.
Puerto Rico	40 min & 45 avg	18	resident	18	no charge	H.S. Equiv. Dipl.
Virgin Islands	40 min & 45 avg	18	none[1]	17	$20.00	H. S. Dipl.

[1] Jurisdictional requirements on exceptions and limitations.
[2] Jurisdictional requirements on credential-related and other fees.

The GED at a Glance			
TEST	**NO. OF QUESTIONS**	**MINUTES TO TAKE TEST**	**SUBJECT AREAS COVERED**
TEST 1 Language Arts: Writing			
Part I	50	75	Part I deals with sentence structure, usage, mechanics, and organization.
Part II	One essay	45	Part II is an essay of about 250 words.
TEST 2 Social Studies	50	70	History, geography, economics, and political science.
TEST 3 Science	50	80	Life science (biology), physical science (chemistry and physics), and Earth and space science.
TEST 4 Language Arts: Reading	40	65	Literary fiction and nonfiction prose.
TEST 5 Mathematics			Arithmetic (measurement, number relationships, and data analysis), algebra, and geometry.
Part 1	25	45	Part 1 allows the use of a calculator.
Part 2	25	45	Part 2 does not allow the use of a calculator.

Total Testing Time: 7 hours, 5 minutes

REA's Test Prep Books Are The Best!

(a sample of the <u>hundreds of letters</u> REA receives each year)

" I am writing to congratulate you on preparing an exceptional study guide. In five years of teaching this course I have never encountered a more thorough, comprehensive, concise and realistic preparation for this examination. "
Teacher, Davie, FL

" I have found your publications, *The Best Test Preparation...*, to be exactly that. "
Teacher, Aptos, CA

" I used your *CLEP Introductory Sociology* book and rank it 99% — thank you! "
Student, Jerusalem, Israel

" Your *GMAT* book greatly helped me on the test. Thank you. "
Student, Oxford, OH

" I recently got the *French SAT II* Exam book from REA. I congratulate you on first-rate French practice tests. "
Instructor, Los Angeles, CA

" Your *AP English Literature and Composition* book is most impressive. "
Student, Montgomery, AL

" The REA *LSAT* Test Preparation guide is a winner! "
Instructor, Spartanburg, SC

(more on front page)